SOME ASPECTS OF AGRICULTURAL DEVELOPMENT IN MAHARASHTRA

SOME ASPECTS OF AGRICULTURAL DEVELOPMENT IN MAHARASHTRA

V.B. BHISE

SERIALS PUBLICATIONS
NEW DELHI

First Published - 2007

ISBN : 81-8387-070-8

Published by

SERIALS PUBLICATIONS
4830/24, Prahlad Street, Ansari Road
Darya Ganj, New Delhi-110002 (India)
Phone : 23245225. Fax : 91-11-23272135
E-mail: serials@satyam.net.in

CONTENTS

PREFACE

I have great pleasure in bringing out this volume containing articles on different aspects of agricultural development in Maharashtra State. The artilces included in this volume are the part of doctoral research works carrried out by the contributors under my supervision at the Dept. of Economics, Dr. Babasaheb Ambedkar Marathwada University, Aurangabad. I am taking the opportunity to publish doctoral research work of the contributors in a book form for the benefit of the teachers and research students in the area of agricultural economics.

Considering the overall economic development, Maharashtra state has been one of the leading states in India. But so far as agriculture sector is concerned, it has not shown much inpressive progress due to some constaints and problem. Some of these problems are deal with in this volume. The issue of relationship between growth and instability in Indian agriculture was widely discussed particularly after the advent of new technology in agriculture. This aspects is studied by R.B. Kanhere and V.B. Bhise. They have discussed the issue of relationship between growth and instability in case of major oilseed crops grown in Maharashtra state . The have worked out annual compound growth rates of area, yield and production of individual oilseed crops at district level to study the performace of the crops. Further, they used different measures for guaging instability in oilseed crop economy of the state.

This volume contains an article by K.N. Rathod which explains the economic position of banjara people in Maharashtra State. This article gives a brief account of historical background of banjaras and presents the results of a survey conducted in thirteen districts of the state where banjara people are largely concentrated.

There is no doubt that cotton and sugarcane are the main commercial crops of the state. Farmers in the state have been largely dependent on these crops. K.K. Patil's article analyses the performance of the latter commercial crop in the state and concludes that this crop has made tremendous progress in the western part of the Maharashtra State.

M.S. Sable's article discusses the acreage response of commercial crops in a backward region of the state . By using Nerlovian partial adjustment model he tried to identify the main factors that govern acreage allocation decisions of the farmers.

It is the well known fact that irrigation is a key input for agricultural development of any region or State. A.N. Pawar's first article explains the development of irrigation sector after independence in the state . His second article analyses the impact of irrigation on development of agriculture by making a case study of one backward district of the Maharashtra State. It is held that inputs like irrigation and chemical fertilizers are excessively used in agricultural production in the country. D.P. Takale's article deals with this issue by conducting a sample study in the Purna irrigation project to estimate resource productivities in the study area. Irrigation was found to be the most important input contributing to crop production. This study indicated a high degree of resource use inefficiency in crop production in the study area.

Agriculture in Maharashtra faces several risks and uncertainties and price uncertainties is one of them . S.R. Takle in his article examined in detail the seasional and long-term variations in prices and market arrivals of main agricultural commodities in the state and made useful suggestions to minimize price uncertainly faced by the farming community in the State.

Maharashtra state occupies top ranks in sugar and milk production in the country. These activities are mostly developed in co-operative sector in the state. B.D. Khandare's article discussed the development of co-operation sugar industry in Maharashtra and also examined sugarcane price policy of the central and State Government's. S.S. Ambhore's article examines the performance of co-operative dairy industry in the State by considering the important parameters of dairy development. Both the authors have made useful suggestions for the development of co-operative sugar and dariy sectors in the State.

I would like to thank the contributors to this volume for submiting their articles well in time. I would like to thank Serials Publications, Delhi for aggreaing to publish this volume speedily. Finally, I thank Kishor Shelar for typing this manuscript.

—V.B. Bhise

LIST OF CONTRIBUTORS

R.B. Kanhere, Joint Director, Higher Education, Aurangabad.

V.B. Bhise, Professor Department of Economics, Dr. Babasaheb Ambedkar Marathwada University, Aurangabad.

K.N. Rathod, Reader and Head, Department of Economics Vasantrao Naik Mahavidyalaya, Aurangabad.

K.K. Patil, Reader and Research Guide, Late Sow. Kamaltai Jamkar Mahila Mahavidyalaya, Parbhani (M.S.)

M.S. Sable, Reader and Head, Department of Economic, Dnyonopasak College, Jintur Dist. Parbhani (M.S.)

A.N. Pawar, Reader in Economic, K.M.J. College, Washi-Osmanabad (M.S.)

D.P. Takale, Reader and Head Department of Economics, Lal Bahadur Shastri Mahavidyalaya, Partur Dist. Jalna (M.S.)

S.R. Takle, Principal and Head, Department of Economics, Narayanrdo Waghmare College, AK. Balapur Dist. Hingoli (M.S.)

B.D. Khandare, Principal, Swami Vivekanand College, Mantha Dist. Jalna (M.S.)

S.S. Ambhore, Department of Economics, Pandit Nehru College, Aurangabad.

R.M. Choudhari, Head Department of Economics, Smt. Sushila Devi Deshmukh Mahila Mahavidyalaya, Latur (M.S.)

R.B. Kanhere & V.B. Bhise

GROWTH AND INSTABILITY IN OILSEED CROP ECONOMY OF MAHARASHTRA

1

INTRODUCTION

In the present global scenario, high growth with more stability carries much importance in the process of agricultural development. Sectoral priorities undergo a change in the process of economic development. In India, as in other developing countries, the economy is diversifying at the macro level with the secondary and tertiary sectors becoming progressively more important in terms of contribution to the national income. Gross value added in Indian agriculture as percentage of Gross Domestic Product (GDP) has reduced to 24.7 percent in 1992-93 from 50.1 percent in 1950-51. Agriculture has grown at the rate about 2.2 percent per annum at a rate that is quite close to the rate of growth of the country's population[1]. At all India level, GDP generated in agriculture sector increased at the rate of 2.16 percent for the period 1968-69 to 1981-82 and 3.28 percent for the period 1981-82 to 1991-92. There is an emergence of significant acceleration in aggregate production and productivity growth in Indian agriculture in the recent years[2].

POSITION OF OILSEED CROPS

India has achieved a comfortable position in the production of cereals. After having attained this, Indian agriculture has made yet another stride by achieving a quantum jump in the production of oilseeds.

India is one of the largest producers of oilseeds in the world with about 25 million hectares of land under oilseed cultivation. India accounts for ten percent of the global oilseed production but uses twenty percent of world land under oilseeds to produce it[3].

Groundnut, sesamum, linseed, sunflower, safflower, rapeseed and mustard, castorseed, nigerseed and soyabean are the important oilseed crops produced in India. It is interesting to note that in respect of several oilseed crops, India either ranks first (groundnut, sesamum, castorseed, linseed and safflower) or second (rapeseed) or third (sunflower) in the world in terms of area under cultivation but its share in the output is comparatively less. Production of oilseeds has increased from 6.2 million tonnes in 1950-51 to 21.4 million tonnes in 1994-95. It is further likely to touch a record level of 24.1 million tonnes in 1996-97. It is interesting to note that substantial contribution to the oilseed production has come from those states which do not have a comparative advantage in the production of cereals[4]. The share of area under oilseed crops showed perceptible decline in the green revolution belt located in the North Western region but the shift was in favour of the oilseed crops at the cost of coarse cereals in the central and southern regions[5].

NEED OF THE STUDY

The issue of growth and instability in agriculture is perhaps, most widely studied. Sen (1971) in his pioneering work at all India level observed that the fluctuations as observed by peaks and troughs were showing markedly divergent trends especially when the production growth comes through area expansion. He stresses on yield fluctuations[6].

Hazell holds the view that production instability is an inevitable consequence of rapid agricultural growth and there is little that can be effectively done about it. To the extent, instability is caused by uncontrollable natural factors. Its intensity can be progressively reduced by controlling the factors that cause instability[7].

Mehra S. (1981) argued in her paper that increase in production has accompanied by increased production instability. Much of this increased instability could be attributed to the widespread adoption of the improved seed/fertilizer intensive technologies since the mid 1960s. Alongwith this, it can also be attributed to factors such as higher sensitivity of yield from the HYV technologies to weather and diseases. Variations in input use inducted by frequent changes and supply restrictions affecting yield etc[8]

However, some of the recent studies, Ray (1983), Dhawan (1987, 1988) and Mahendra Dev (1987) differ from the earlier studies of Mehra (1981) and Hazell (1982) which emphasized that instability is a consequence of growth.

Maharashtra State has different pattern of growth than the high growth states like Punjab, Haryana, Tamil Nadu etc. In the case of Maharashtra, Nadkarni (1971) had earlier observed that there is strong negative relationship between the level of instability and growth rate in yield across crops. This was mainly due to the slow growth crops showing high instability[9]. This was true in the pretechnology period when the new forces of growth had not entered. Recently, Deshpande R.S. (1988) observed that Maharashtra State has high instability with low growth at least up to 1980-81[10]. Mitra A.K. (1990) noted that whatever increases in the rates of growth in crop production, especially foodgrains, have been achieved are seen to be positively associated with instability in production brought about mainly by relative high yield instability[11].

Many studies relating to oilseed crops in Maharashtra have come up. Notable studies are those of Ashturkar (1988), Hedgire and Acharya (1988), Korde and Mahalle (1990), Waghmare and Katepattewar (1993), Rahane and Joshi (1993), Sale, Pawar and Satte (1993). These studies, in general, had an impression that the growth in oilseed crops in Maharashtra is satisfactory but it suffers on account of high instability.

As far as oilseed crop studies in Maharashtra are concerned, it can be said that many of the studies are restricted to the state and/or regional level. In brief, no one has, so far, studied in detail the growth and instability in all major oilseed crops in Maharashtra at the district level. The present study attempts to bridge this gap. So it is wroth to study the growth and instability in major oilseed crops in Maharashtra at the regional and/or district level.

THE PRESENT PAPER ATTEMPTS TO STUDY

1. The growth in area, production and yield of major oilseed crops in Maharashtra at the regional and district levels;
2. The instability in yields of major oilseed crops in Maharashtra at the regional and district levels, and
3. The nexus between growth and instability in yields or major oilseed crops at the Maharashtra state level.

RESEARCH METHODOLOGY

This study pertains to the oilseed crop economy of Maharashtra. It attempts to cover groundnut, sesamum, linseed, safflower, sunflower, nigerseed and castorseed as major oilseed crops of the state. The time span of 25 years from 1968-69 to 1992-93 is considered for the study. Four regions viz. Konkan region, Western Maharashtra region, Marathwada region and Vidarbha region are considered in the study. 26 districts (position before 1982) are considered for the study.

Data about area, production and yield of major oilseed crops required for the study was collected from the published sources such as:

1. Season and Crop Reports of Maharashtra
2. Districtwise General Statistical Information of Agriculture Department (Epitome of Agriculture) in Maharashtra, Part II;
3. Economic Survey of Maharashtra;
4. Office of the Directorate of Agriculture, Pune.

To calculate growth rates of area, production and yield of major oilseed crops, the exponential function of the form y = abt was employed. For operational purpose, it becomes a semilog function of the form: log y=a+bt, where y stands for area/production/yield. a, b are constants and t the time span. (b-1)´100 gives percent compound annual growth rate. Standard error of the growth rate was worked out by using the formula No.1 given in the appendix. Significance of the growth rates was tested by using student's "t" test, t = r/SE(r) where t = student's 't' distribution with (n-2) degrees of freedom, r = compound annual growth rate and SE(r) = standard error of growth rate. Instability in yield of major oilseed crops was estimated by using four different measures viz., coefficient of variation around the trend line (CVTL), coefficient of variation (CV), probability of crop failure [p(CF)] and Crop Loss Ratio (CLR). The formulae for these measures are given in the appendix. Cropwise yield growth and instability nexus was viewed at the state level by classifying districts into low, medium and high growth and instability in yields.

GROWTH PERFORMANCE OF MAJOR OILSEED CROPS IN MAHARASHTRA: POSITION OF OILSEEDS IN MAHARASHTRA

Maharashtra is the third largest state in terms of geographical area, population and per capita income (Rs. 15244 at current prices for 1995-96). Though Maharashtra is considered progressive state, it is deficit when one considers agriculture. The relative share of agriculture at the national level has declined from 50.1 percent in 1950-51 to 24.7 percent in 1992-93 whereas it has declined to 21.2 percent in 1994-95 in Maharashtra.

As far as oilseed crops are concerned, Maharashtra ranks first in the production of safflower and linseed. The state has retained its fifth position in the production of total oilseeds over a period of time. The state has 9.8 percent area under oilseed crops compared to the total area under oilseed crops in India and it has contributed around 8.64 percent to the total production in 1993-94. Areawise relative position of the important oilseeds in Maharashtra for 1990-93 was as follows: Groundnut (23.49 percent), sesamum (11.43 percent), safflower (19.40 percent), sunflower (17.72 percent), linseed (7.07 percent), nigerseed (3.34 percent) and castorseed (0.23 percent) respectively. Groundnut is the leading oilseed crop which is followed by sunflower, safflower, sesamum, linseed and nigerseed in the state.

CROP WISE GROWTH PERFORMANCE OF OILSEEDS

Cropwise annual compound growth rates of area, production and yield for the period 1968-69 to 1992-93 are depicted in table 1. It can be observed from the table that the total oilseeds in Maharashtra had 5.32

percent annual compound production growth rate, 3.88 percent yield growth rate and 1.88 percent area growth rate for 1968-69 to 1992-93. All these growth rates were found to be significant at one percent level.

Table 1
Annual Compound Growth Rates of Oilseed Crops in Maharashtra (Period 1968-69 to 1992-93).

(Percent per annum)

Sr. No.	Name of the crop	Area	Production	Yield
1.	Groundnut	-1.99***	-0.01	2.07***
		(0.23)	(0.69)	(0.67)
2.	Sesamum	3.60***	5.17***	1.88***
		(0.66)	(0.78)	(0.47)
3.	Linseed	-0.62	0.97	-0.62
		(0.52)	(2.05)	(0.52)
4.	Safflower	1.38***	4.68***	2.64***
		(0.46)	(1.00)	(0.63)
5.	Sunflower	17.36***	14.30***	-1.82
		(4.39)	(3.84)	(2.15)
6.	Nigerseed	-14.82***	15.93***	1.17***
		(1.71)	(2.08)	(0.66)
7.	Castorseed	1.78***	3.06***	2.03***
		(0.63)	(1.89)	(0.98)
8.	Total Oilseeds	1.88***	5.32***	3.88***
		(0.29)	(0.76)	(0.62)

Note: 1. # Compound growth rates refer to the period 1980-81 to 1992-93
2. ***Significant at one percent level.
3. **Significant at five percent level.
4. *Significant at ten percent level.
5. Growth rates are based on the data collected from
(a) Season and Crop Report of Maharashtra and
(b) Districtwise General Statistical Information of Agricultural Department (Epitome of Agriculture) in Maharashtra, Part II.

Production growth rates for sesamum, safflower, sunflower, nigerseed were significant at one percent level while these growth rates were not significant for groundnut and linseed. The production of groundnut and linseed was almost stagnant over the period from 1968-69 to 1992-93. These two crops experienced negative area growth rates where the fall in the area under groundnut was significant at one percent level. The area growth rates were positive and significant at one percent level for sesamum, safflower, sunflower, nigerseed, castor seed and total oilseeds as a category. Yield growth rates were positive and significant at one percent level for groundnut, sesamum, safflower and total oilseeds and significant at 5 percent level for nigerseed and castorseed crops. Negative and insignificant growth rates were observed in the case of linseed and sunflower crops. Periodwise consideration of these crops reveal that the production growth of the major oilseeds has come mainly through area expansion in the green revolution period i.e. from 1968-

69 to 1978-79 while there was no spectacular fall in the area under oilseed crops cultivation in the post-green revolution period i.e. 1979-80 to 1992-93.

REGIONWISE GROWTH PERFORMANCE OF OILSEEDS

Regionwise growth rates of area, production and yield of total oilseed crops in Maharashtra are shown in Table 2.

Table 2
Regionwise Growth Rates of Area, Production and Yield of Total Oilseed Crops in Maharashtra during 1968-69 to 1992-93

(Percent per annum)

Sr. No.	*Region*	*Area*	*Production*	*Yield*
1.	Konkan	0.05	13.32***	13.26***
		(0.97)	(1.71)	(1.65)
2.	Western Maharashtra	1.75***	4.60***	2.79***
		(0.30)	(0.81)	(0.68)
3.	Marathwada	2.08***	6.91***	4.70***
		(0.41)	(1.30)	(1.11)
4.	Vidarbha	2.30***	4.39***	2.04***
		(0.34)	(0.73)	(0.54)
5.	Maharashtra	1.88***	5.32***	3.38***
		(0.29)	(0.76)	(0.62)

Source: As mentioned in Table 1.

It is clear from the table that the production growth rates were positive and significant at one percent level in all the regions alongwith the state level for the period 1968-69 to 1992-93, though we had bigger growth rate for the Konkan region. It carries less weightage as production of oilseed crops in this region is very negligible. Therefore, it can be stated that production growth rate for the Marathwada region was the highest which was to the tune of 6.91 percent in the period from 1968-69 to 1992-93. Further, one can note that the yield growth rates were also positive and significant in all the regions along with the State in Maharashtra. It can also be observed that the area growth rates were positive and significant in the regions viz., Western Maharashtra, Marathwada and Vidarbha. Konkan region had insignificant area growth rate indicating almost stagnancy in the use of area for oilseed crop cultivation.

Considering area, production and yield growth rates for total oilseeds in Maharashtra during the period 1968-69 to 1992-93, one may conclude that Western Maharashtra, Marathwada and Vidarbha regions are the important regions in the production of oilseed crops.

DISTRICTWISE GROWTH PERFORMANCE OF OILSEEDS

Districtwise growth rates of area, production and yield of total oilseed crops in Maharashtra during the period of 1968-69 to 1992-93 are depicted in Table 3 which reveal that area, production and yield growth rates were positive and significant at one percent level in the districts such as Ahmednagar, Pune, Solapur, Satara, Aurangabad, Parbhani, Buldhana, Wardha and Nagpur. Beed district was having area and production growth rates positive and significant at one percent level but the yield growth rate was positive and significant at five percent level. Jalgaon, Kolhapur, Nanded, Akola, Amravati and Chandrapur districts recorded positive and highly significant (one percent level) area and production growth rates but the yield growth rates in these districts were insignificant. Whereas Sangli, Osmanabad, Bhandara, Ratnagiri, Thane and Nasik districts were having positive and significant production and yield growth rates but have insignificant area growth rates whereas only Yavatmal district has significant negative growth rate during 1968-69 to 1992-93. The production growth rate was negative and insignificant in Yavatmal district only while it was positive and insignificant in Dhule district. Only Jalgaon and Akola districts were with negative and insignificant yield growth rates.

Table 3
Districtwise Growth Rates of Area, Production and Yield of Total Oilseed Crops in Maharashtra (Period 1968-69 to 1992-93)

(Percent per annum)

Sr. No.	*District*	*Area*	*Production*	*Yield*
1.	Thane	0.73	15.08***	1.11*
		(1.06)	(1.93)	(0.77)
2.	Raigad	0.80	6.05***	6.43
		(2.01)	(1.84)	(40.57)
3.	Ratnagiri	(-0) 0.34	14.65***	15.02***
		(1.03)	(1.92)	(1.82)
4.	Nasik	(-) 0.31	1.64***	1.97***
		(0.37)	(1.20)	(1.05)
5.	Dhule	(-) 0.15	0.76	0.92
		(0.25)	(1.99)	(1.13)
6.	Jalgaon	3.65***	3.29***	(-) 0.08
		(0.58)	(1.17)	(0.82)
7.	Ahmednagar	3.28***	11.87***	8.82***
		(0.62)	(1.78)	(1.81)
8.	Pune	2.60***	8.23***	5.60***
		(0.36)	(1.06)	(0.94)
9.	Solapur	2.37***	8.86***	6.35***
		(0.67)	(2.14)	(1.65)
10.	Satara	0.90***	3.44***	2.52***
		(0.30)	(0.89)	(0.90)
11.	Sangli	(-) 0.06	3.91***	3.98***
		(0.66)	(1.24)	(1.06)
12.	Kolhapur	2.16***	5.26***	0.86
		(0.41)	(0.75)	(1.23)
13.	Aurangabad	1.66***	11.27***	8.02***
		(0.56)	(1.83)	(1.79)

Contd....

Sr. No.	District	Area	Production	Yield
14.	Parbhani	2.05*** (0.53)	8.85*** (1.43)	6.69*** (1.47)
15.	Beed	1.99*** (0.51)	7.15*** (1.93)	3.47*** (1.80)
16.	Nanded	6.06*** (1.19)	5.86*** (1.74)	0.67 (1.00)
17.	Osmanabad	0.73 (0.63)	6.06*** (1.63)	3.76*** (1.21)
18.	Buldhana	3.06*** (0.53)	6.89*** (1.01)	3.22*** (1.05)
19.	Akola	2.22*** (0.54)	2.01*** (1.01)	(-) 0.39 (0.78)
20.	Amravati	2.34*** (0.60)	2.70*** (0.99)	0.23 (0.74)
21.	Yavatmal	(-) 1.12*** (0.44)	(-) 0.49 (0.88)	0.50 (0.75)
22.	Wardha	4.66*** (0.52)	7.42*** (0.97)	2.12*** (0.84)
23.	Nagpur	4.28*** (0.54)	7.72*** (1.11)	2.77*** (0.72)
24.	Bhandara	(-) 0.49 (0.54)	4.34*** (1.53)	3.92** (1.58)
25.	Chandrapur	1.28*** (0.26)	3.43*** (0.89)	0.99 (0.94)
	Maharashtra State	1.88*** (0.29)	5.32*** (0.76)	3.38*** (0.62)

Source: As shown in Table 1.

It can also be inferred from the table that production growth has come mainly through significant area expansion in Jalgaon, Kolhapur, Nanded, Akola, Amravati and Chandrapur districts during 1968-69 to 1992-93 while significant production growth was the result of significant yield growth in Ratnagiri, Nasik, Sangli, Osmanabad and Bhandara districts. In rest of the districts, the production growth was a combined effect of area and yield growth over the period under consideration. Considering production growth as a criterion, one may state that Ahmednagar, Pune, Solapur, Kolhapur, Aurangabad, Parbhani, Beed, Nanded, Osmanabad, Buldhana, Wardha and Nagpur are the important oilseed producing districts in Maharashtra. The contribution of districts from the Konkan region has remained negligible though these districts have higher production growth rates. So, they are considered less important.

YIELD INSTABILITY IN THE MAJOR OILSEED CROPS IN MAHARASHTRA

Yield instability in the major oilseed crops in Maharashtra has been considered to study instability aspect. The instability was worked out by employing four measures viz., Coefficient of Variation around the

Trend Line (CVTL), coefficient of variation (CV), Probability of Crop Failure [P(CF)], and Crop Loss Ratio (CLR). These measures are expressed in percentages. So, they are comparable.

CROPWISE POSITION OF YIELD INSTABILITY

Cropwise yield instability in oilseed crops in Maharashtra during 1968-69 to 1992-93 is presented in Table 4.

Table 4
State Level Yield Instability of Major Oilseed Crops in Maharashtra During 1968-69 to 1992-93

Crop	*Yield Instability*			
	CVTL	*CV*	*P(CF)*	*CLR (-)*
Groundnut	18.76	22.97	28	6.04
Sesamum	17.53	21.86	20	4.38
Linseed	15.76	15.89	24	4.61
Safflower	24.13	28.73	36	7.16
Sunflower	26.81	25.89	34.46	9.48
Nigerseed	25.42	26.81	28	8.56
Castorseed	39.67	39.33	28	7.95
Total Oilseeds	18.10	28.59	20	5.15

Note: 1. The data presented in this table is based on yield of major oilseed crops in Maharashtra.
2. Figures for this crop refer to 1980-81 to 1992-93.

Source: 1. Season and Crop Reports of Maharashtra, various issues.
2. Districtwise General Statistical Information of Agriculture Department (Epitome of Agriculture) of Maharashtra – various issues.

It reveals that the highest yield instability was observed in the case of castorseed crop considering coefficient of variation around the trend line (CVTL) and coefficient of variation (CV) as the measures of instability. Whereas the lowest instability was recorded by the linseed crop during 1968-69 to 1992-93. Sunflower, nigerseed, safflower, groundnut, sesamum and linseed crops come in a descending order of instability in the state when CVTL is considered as a measure of instability. However, the instability ranking changes when other measures are considered. The total oilseeds category shows that instability as per CVTL was 18.10 percent, it was 28.59 percent as per CV and 20 percent as per probability of crop failure. Crop loss ratio for total oilseeds was 5.15. Probability of crop failure for individual crops from the oilseeds was recorded higher than the probability of crop failure for total oilseed crops category during this period under study. In brief, groundnut, sesamum and linseed crops along with the total oilseed crop category had comparatively lower yield instability while it tends to be higher for safflower, sunflower, nigerseed and castorseed crops in Maharashtra during 1968-69 to 1992-93. Safflower and sunflower, though important oilseed crops of the state, safflower on account of comparatively higher instability is a point of concern for the oilseed crop economy of Maharashtra.

REGIONWISE POSITION OF YIELD INSTABILITY OF TOTAL OILSEED CROPS IN MAHARASHTRA

The regionwise yield instability in total oilseed crops in Maharashtra during 1968-69 to 1992-93 is given in Table 5.

Table 5
State Level Yield Instability of Total Oilseed Crops in Maharashtra During 1968-69 to 1992-93

Regions	*Yield Instability*			
	CVTL	*CV*	*P(CF)*	*CLR (-)*
Konkan	34.87	82.19	44	9.47
Western Maharashtra	18.13	25.13	20	5.03
Marathwada	40.55	46.26	32	10.60
Vidarbha	19.15	24.43	32	6.35
Maharashtra state	18.10	28.59	20	5.15

Source: As mentioned in Table 4.

It shows that higher yield instability was noticed in the Marathwada and Konkan regions considering CVTL and CV as the measures of yield instability on one hand and comparatively lower yield instability in the Western Maharashtra and Vidarbha regions of the Maharashtra state. Considering CVTL and P(CF) as the measures of yield instability it can be noted that yield instability recorded by the individual regions was comparatively higher than the state level yield instability. One more thing, yield instability measured in terms of CV proved to be comparatively higher than the yield instability measured in terms of CVTL in all the regions. Probability of crop failure was the highest in Konkan region while it was the lowest in the Western Maharashtra region. It was observed as the same in the Marathwada and Vidarbha regions. In brief, Konkan region is not important in the oilseed crop production and many districts of the Vidarbha region do not seem to be important in the oilseed crop production. Further that Marathwada region has high degree of yield instability on one hand and Western Maharashtra region has low yield instability on the other hand.

DISTRICTWISE YIELD INSTABILITY OF TOTAL OILSEED CROPS IN MAHARASHTRA

Districtwise position of yield instability of total oilseed crops in Maharashtra during 1968-69 to 1992-93 is depicted in Table 6. It can be seen from the table that the yield instability of individual districts was comparatively higher than the state level yield instability measured in terms of CVTL for all the districts in the State of Maharashtra. Inter-district comparison of yield instability further reveals that districts from the Konkan and Marathwada region contribute comparatively higher degree of instability than the districts from the Western Maharashtra and Vidarbha regions. If yield instability below 25 percent is considered as low instability, yield instability between 25 percent to 35 percent as medium instability and above 35 percent as high instability. Then the conclusion derived, considering CVTL as a measure of instability, tells us that Pune, Sangli, Amravati and Yavatmal districts along with the state as a whole has low instability while Raigad, Ratnagiri, Ahmednagar, Solapur, Aurangabad, Parbhani, Beed, Bhandara and Chandrapur districts have high yield instability during 1968-69 to 1992-93. Rest of the districts have medium degree of instability for oilseed crops.

Table 6
Districtwise Yield Instability of Total Oilseed Crops in Maharashtra During 1968-69 to 1992-93

Sr. No.	*District*	*Yield Instability*			
		CVTL	*CV*	*P(CF)*	*CLR (-)*
1.	Thane	27.17	26.60	28	6.19
2.	Raigad	54.83	59.51	32	9.47
3.	Ratnagiri	38.61	94.31	44	9.50
4.	Nasik	28.63	29.13	28	8.84
5.	Dhule	32.49	31.39	28	10.13
6.	Jalgaon	27.37	25.97	28	8.45
7.	Ahmednagar	48.18	59.81	40	12.06
8.	Pune	23.05	41.21	24	6.19
9.	Solapur	41.89	51.99	36	12.39
10.	Satara	25.15	28.46	28	6.28
11.	Sangli	24.15	33.02	20	5.39
12.	Kolhapur	30.14	29.83	20	7.05
13.	Aurangabad	47.42	54.05	40	12.16
14.	Parbhani	47.62	55.97	44	12.30
15.	Beed	49.74	48.32	32	11.65
16.	Nanded	33.08	31.48	40	10.38
17.	Osmanabad	32.38	38.67	32	8.79
18.	Buldhana	31.14	36.48	32	8.69
19.	Akola	26.62	25.59	32	8.49
20.	Amravati	23.85	22.70	32	7.45
21.	Yavatmal	24.75	23.78	36	8.68
22.	Wardha	27.83	31.23	40	8.74
23.	Nagpur	26.35	33.65	40	7.76
24.	Bhandara	52.98	55.42	32	11.15
25.	Chandrapur	41.69	40.26	40	7.97
	Maharashtra State	18.10	28.59	20	5.15

Source: As mentioned in Table 4.

NEXUS BETWEEN YIELD GROWTH AND YIELD INSTABILITY OF MAJOR OILSEED CROPS IN MAHARASHTRA

For considering the nexus between yield growth and yield instability of oilseed crops in Maharashtra during 1968-69 to 1992-93, the growth rates are grouped into four categories as high, medium, low and negative. Growth rates above three percent are considered as high, growth rates between two percent to three percent as medium, growth rates between zero to two percent as low and growth rates less than zero are considered as negative. The yield instability measured in terms of CVTL is considered for the nexus. It is grouped into three categories viz., low, medium, and high. CVTL which is less than 25 percent is considered as belonging to low category, CVTL between 25 percent to 35 percent is considered as medium instability and CVTL above 35 percent constitute high instability.

CROPWISE NEXUS BETWEEN YIELD GROWTH AND YIELD INSTABILITY

Table 7 presents the oilseed crops classified according to the growth rates and instability of yield in Maharashtra during 1968-69 to 1992-93. It is clear from the table that total oilseeds as a category forms the most preferred situation of high growth with low instability of yield expressing much scope for advancing oilseed crops in the Maharashtra State.

Table 7
Oilseed Crops Classified According to the Growth Rates and Instability in Yield in Maharashtra During 1968-69 to 1992-93

Growth	*Yield Instability (CVTL)*		
	Low (Below 25%)	*Medium (25 to 35%)*	*High (above 35 %)*
High (above 3%)	Total oilseeds	-	-
Medium (2 to 3%)	Groundnut Safflower	-	Castorseed
Low (0 to 2%)	Sesamum Oilseeds	Nigerseed	-
Negative	Linseed	-	-

Note: Yield growth rates and instability (CVTL) for the crops was worked out on the basis of the data collected from.

(a) Season and Crop Reports of Maharashtra–Various issues.

(b) Districtwise General Statistical Information of the Agriculture Department (Epitome of Agriculture) Maharashtra–various issues.

Groundnut and safflower crops have medium yield growth with low instability indicating their better position amongst the oilseed crops. Castorseed also had medium growth but the yield instability happens to be high for this crop. Sesamum and linseed crops in the State could not take advantage of low instability during 1968-69 to 1992-93. In general, yield instability for majority of the oilseed crops observed to be low followed with varying levels of yield growth in the Maharashtra State. So, efforts should be concentrated on enhancing the oilseed production of different oilseed crops by following area expansion as well as yield improvement which will be in the interest of the oilseed crop economy of Maharashtra.

REGIONWISE NEXUS BETWEEN YIELD GROWTH AND YIELD INSTABILITY

Regions classified according to the growth rates and instability in yield of total oilseed crops in Maharashtra during 1968-69 to 1992-93 are presented in Table 8.

Table 8
Regions Classified According to the Growth Rates and Instability in Yield of Total Oilseed Crops in Maharashtra During 1968-69 to 1992-93.

Growth	*Yield Instability (CVTL)*		
	Low (Below 25%)	*Medium (25 to 35%)*	*High (above 35 %)*
High (above 3%)	Maharashtra State	Konkan region	Marathwada region
Medium (2 to 3%)	Western Maharashtra region Vidarbha region	-	-
Low (0 to 2%)	-	-	-
Negative	-	-	-

Note: As mentioned in Table 7.

It is clear from the table that Maharashtra State falls in the most preferred combination of high growth with low instability for total oilseed crops. But the regionwise classification reveals that Western Maharashtra and Vidarbha regions have medium growth followed by low instability measured in terms of CVTL. Konkan region has high growth which is followed by medium yield instability. Marathwada region has high growth with high instability. Konkan region carries less importance in the analysis. So, this enables to state that efforts must be resorted to reduce the degree of yield instability in the Marathwada region on one hand and to take the advantage of low yield instability in the Western Maharashtra and Vidarbha regions by boosting the yield growth on the other. These two fold measures can improve the oilseed crop economy of Maharashtra.

DISTRICTWISE NEXUS BETWEEN YIELD GROWTH AND YIELD INSTABILITY

Districts classified according to the growth rates and instability in yield of total oilseed crops in Maharashtra during 1968-69 to 1992-93 is depicted in Table 9. It is clear from the table that Pune and Sangli districts along with the Maharashtra State have high growth with low instability. Osmanabad and Buldhana districts have high growth with medium instability while Raigad, Ratnagiri, Ahmednagar, Solapur, Aurangabad, Parbhani, Beed and Bhandara districts have high growth with high instability. Satara, Wardha and Nagpur districts experienced medium growth with medium instability. Amravati and Yavatmal districts cannot take advantage of low instability for improving growth. Jalgaon and Akola districts registered negative yield growth with medium instability in the State while Chandrapur district faces high instability with low growth which is perhaps not a preferred combination. Of course, one must pay attention to the point that this position vary from crop to crop and from period to period even in the Maharashtra State itself.

Table 9
Districts Classified According to the Growth Rates and Instability in Yield of Total Oilseed Crops in Maharashtra During 1968-69 to 1992-93.

Growth	*Yield Instability (CVTL)*		
	Low (Below 25%)	*Medium (25 to 35%)*	*High (above 35 %)*
High (above 3%)	Pune Sangli, M.S.	Osmanabad Buldhana	Raigad Ratnagiri, Ahmednagar, Solapur, Aurangabad, Parbhani, Beed, Bhandara
Medium (2 to 3%)	-	Satara, Wardha Nagpur	-
Low (0 to 2%)	Amravati, Yavatmal	Thane, Nasik, Dhule, Kolhapur, Nanded, Jalgaon, Akola	Chandrapur
Negative			

Note: As mentioned in Table 7.

MAIN CONCLUSION AND POLICY IMPLICATIONS

Total oilseed crops demonstrated significant production growth of 5.32 percent during the period 1968-69 to 1992-93 where yield growth (3.38%) was comparatively higher than the area growth (1.88%) in Maharashtra. All these growth rates were significant at one percent level. Area expansion was more prominent in the sesamum, safflower and sunflower crops while groundnut and linseed crops lost some area under cultivation. Regionwise analysis revealed that Western Maharashtra and Marathwada regions proved to be important producers of oilseed crops in Maharashtra while Konkan region do not constitute an important producer of oilseed crops. Jalgaon, Ahmednagar, Pune, Solapur, Satara, Sangli, Kolhapur, Aurangabad, Beed, Osmanabad,

Nanded, Parbhani, Buldhana, Amravati, Wardha, Nagpur, Bhandara and Chandrapur are the important oilseed producing districts of Maharashtra. It implies that Maharashtra State depicts satisfactory production growth performance of total oilseeds on account of favourable agro-climatic condition. So, efforts must be concentrated on increasing the oilseed production in the state. Further that more emphasis should be placed on Western Maharashtra and Marathwada regions along with the important oilseed producing districts of the State.

The study also revealed that total oilseed crops have high yield growth with low yield instability during the overall period. Regional analysis showed that Western Maharashtra and Vidarbha regions have moderate yield growth with low yield instability whereas Marathwada and Konkan regions have high yield growth followed by high yield instability. District level analysis helped to note that only Pune and Sangli districts were having high yield growth with low yield instability i.e. the most preferred combination while Raigad, Ratnagiri, Ahmednagar, Solapur, Aurangabad, Parbhani, Beed and Bhandara districts were having high yield growth with high yield instability for total oilseeds in Maharashtra during the period of 25 years under consideration. These results imply that it could be in the interest of the oilseed crop economy of Maharashtra to stress more on the Western Maharashtra region–an important oilseed producing region, where yield instability is low. Further efforts must be taken to reduce the higher degree of instability in the another important oilseed producing region i.e. Marathwada. Simultaneously, yield instability must be lessened from the above mentioned important oilseed producing districts where high yield instability prevailed.

It is well accepted that around thirty three percent of the State area is drought prone and water table in the State is on fast decline on account of less recharging of water in Maharashtra. It has been observed that oilseed crops can be grown satisfactorily in the dryland area. Not only this, but it has been shown that returns are two to five times more in the rainfed conditions. Therefore, it could be imperative to state that farmers in Maharashtra should resort to farming of oilseed crops so as to improve their economic position. It has also been noticed that many dryland farmers grow coarse cereals as a tradition. Recently, it is rightly pointed out by the researchers that coarse cereals can be successfully replaced by oilseed crops. So, the policy implication in this regard is that wherever possible coarse cereals should be replaced by oilseed crops. This will be in the interest of the oilseed crop economy of Maharashtra.

NOTES & REFERENCES

1. Rao V.M. (1994), Agricultural Liberalisation: Some Implications for the Development Policies. Economic and Political Weekly, Vol. 29 (16 and 17).

2. Sawant S.D. & Achuthan C.V. (1995), Agricultural Growth Across Crops and Regions–Emerging Trends and Patterns. Economic and Political Weekly, Vol. 30 (12), pp. A-2 to A-13.
3. Singhal Vikas (1995), Handbook of Indian Agriculture. Vikas Publishing House Pvt. Ltd., Page 48.
4. Acharya S.S. (1993), Oilseeds and Pulses-Price Policy and Production Performance. Indian Journal of Agricultural Economics, Vol. 48 (3), pp. 317-333.
5. Bhalla G.S. and Singh Gurmail (1997), Recent Developments in Indian Agriculture–A State Level Analysis. Economic and Political Weekly, Vol. 32(13), A-2 to A-18.
6. Sen S.R. (1971), Growth and Instability in Indian Agriculture. Firma, K.L. Mukhopadhyay, Calcutta.
7. Ray S.K. (1979), An analysis of Food Production and Grain Stock Potentials in India. International Development Series, Iowa State University, Ames, Iowa, U.S.A.
8. Mehra S. (1981), Instability in Indian Agriculture in the Context of the New Technology. Research Report 25, International Food Policy Research Institute, Washington, D.C.
9. Nadkarni M.V. (1971), Yield Uncertainty in Maharashtra Agriculture, Indian Journal of Agricultural Economics, Vol. 26 (4), pp. 327-333.
10. Deshpande R.S. (1988), Growth and Instability in Maharashtra Agriculture, Artha Vijanya, Vol. 30 (4), pp. 317-339.
11. Mitra A.K. (1990), Agricultural Production in Maharashtra-Growth and Instability in the Context of New Technology, Economic and Political Weekly, Vol. 25 (52), pp. A-146 to A-164.
12. Ashturkar, B.W. (1988), Performance of Oilseed Crops in Maharashtra (Project). Agresco Report 1987-88, p. 1, Marathwada Agricultural University, Parbhani.
13. Hedgire D.N. and Acharya H.S. (1988), Cotton Seed as an Additional Source of Edible Oil. Agresco Report 1987-88, p. 14, Marathwada Agricultural University, Parbhani.
14. Korde M.D. and Mahalle Y.P. (1990), Instability in Productivity of Oilseed Crops Growth in Nagpur Division, Indian Journal of Agricultural Economics, Vol. 43(3), pp. 316.
15. Waghmare P.R. and Katepattewar B.N. (1993), Districtwise Trends in Area, Production and Productivity of Oilseeds in Maharashtra State: 1980-81 to 1989-90, Indian Journal of Agricultural Economics, Vol. 43(3), pp. 425, 426.
16. Rahane R.K. and Joshi G.G. (1993), Growth Rates in Area, Production and Productivity of Some Important Oilseeds and Pulses in Maharashtra, Indian Journal of Agricultural Economics, Vol. 48(3), p. 416.
17. Sale D.L., Pawar J.R. and Sethe V.R. (1993), Performance of Important Oilseeds in Maharashtra–A regionwise Analysis, Indian Journal of Agricultural Economics, Vol. 48(4), pp. 408, 409.

APPENDIX

1. Standard Error of Growth Rate.

$$SE\ (r) = \frac{100 \times B}{\log_{10} e} \sqrt{\frac{\sum_{1}^{n} \log y^2 - \frac{(\sum \log y)^2 - (\log b)^2}{n} - \sum_{1}^{n} (t - \bar{t})^2}{(n-2) \sum_{1}^{n} (t - \bar{t})^2}}$$

Where,

Se(r) = Standard error of compound growth rate

B = r + 1,

$\text{Log}_{10}\ e$ = 0.43429

y = area/production/yield of the crop.

t = time factor, 1 = 1, 2 ......n

n = number of observations.

2. $$\text{CVTL} = \sqrt{\frac{\sum (y_i - \hat{y}_1)^2 / (n-2)}{\bar{y}_i}} \times 100$$

where,

CVTL = Coefficient of variation around the trend line.

$\Sigma(y_1 - \hat{y}_1)^2$ = The sum of the squared residuals of the regression y on time for i[th] crop.

y = yield of the ith crop,

n = number of observations,

$\hat{y}_1$ = Mean yield of the ith crop.

3. $$\text{CV} = \frac{\sigma}{\bar{x}} x100$$

Where CV = Coefficient of variation,

σ = Standard deviation of the crop yield.

$\bar{x}$ = Mean of the crop yield.

4. $$P(CF) = \frac{n(f)}{n} \times 100$$

Where, P(CF) = Probability of crop failure,

$n(f)$ = number of crop failure years,

n = total number of years.

A crop failure year is one in which crop yield is less than ten percent of the trend yield of the crop.

5. $CLR \frac{\sum_{j=1}^{m} d_j}{\sum_{t-1}^{n} y_t}$

where, CLR = Crop loss ratio.

d_j = deviations of actual yield from the yield below ten percent of the trend estimated for the year.

m = number of crop failure years.

y_1 = yield for the total number of years.

n = total number of years.

K.N. Rathod

ECONOMIC POSITION OF BANJARA COMMUNITY IN MAHARASHTRA

2

Banjara people of our country is that part of Indian society which were engaged in trade and carrying occupation on pack bullocks from generation to generation, before modern means of transport like railway came into existence. This occupation is known as their traditional occupation. It is an age-old tradition of our country to know a particular group of people from their traditional occupation. The word 'Banjara' is derived from the original Sanskrit word 'Vanijyam'. The meaning of this word is trade[1]. In former time they were wandering from one part of the country to another part for the purpose of carrying goods and trade. This is the reason why they are known in history as a nomadic tribe.

Modern means of quick transport were not available in the seventeenth century and before. Oxen were mainly used for transport of goods. Other animals like horses, elephants were not in use for transport of goods as they were used for fighting in war and were reserved for the kings and high dignitaries. Raths were used for passenger traffic but they were very expensive. Camels were used in sandy areas. Under such situation oxen were the only important carrier. Under such circumstances, transport of goods on the back of oxen became a profession. In ancient period, this occupation was in the hands of the people who were known as Banjaras. Banjaras were a specialist class of carriers who were moving from one part of the country to another with their families. In ancient period, many European travelers observed the carvans (tandas) moving from one place to another. These carvans numbering 10,000 to 12,000 oxen together and some time 20,000 oxen also[2].

Now-a-days, these Banjaras are not moving as before; their traditional occupation is also gone. They are mainly settled down in hilly track and mostly depend directly and indirectly on agriculture. This nomadic tribe, now-a-days, spread all over the country. Although it is the fact, but they are located in central India in majority, especially in Maharashtra state. It is the fact that northern India is their original home. Before migrating to South they were doing their traditional occupation in the north especially in Gujarath and Marwad regions. Their customs, culture, dress, dialect and even historical record proves their northern origin.

History of Banjara community shows that their migration process from North to South India started when Mohammad Tughlaq, King of Delhi, marched to Deccan (Daulatabad) in the fourteenth century. He employed Banjaras as carriers[3]. According to historical records, it seems that their migration from North to South India began in fourteenth century; but they migrated systematically and in large number with their origin and organization evidently came to Deccan in fourth decade of seventeenth century with Asat Jan sometime called Asaf Khan the Wazir of King Shaha Jahan, when Mughal wanted to step down in South India. In those days Mughal wanted to spread their political power in South India. For this purpose, they had to fight different powers in South. But even in those days, armies had to return even without fighting the battle as the food supply may not reach them in time. There did not exist any designated category to undertake this responsibility of supply of food to army camp. In those circumstances the Mughal made a contract with Banjaras of North and Banjaras migrated in large numbers from North to South India[4].

The aim of Mughal to bright Banjaras with them from North India was politically motivated. They wanted better foodgrain for their army so that they could win the war and settle down politically in South India. But the aim of Banjaras to migrate with Mughal army was not politically motivated. They were looking toward their traditional occupation purely from business point of view. This is the reason that they supplied foodgrain to all political powers in South, namely, Mughal, Maratha, Peshwa and finally British too. Their services were available to any political power on demand. It is interesting to know that in one war whom they supplied foodgrains; in very next war they supplied foodgrains to opposite political power. Who avails their services make no difference for them. It shows their non-political attachment and occupational approach. Supplying foodgrains at the time of war and in time of water interval engage in internal trade were two main aspects of their traditional occupation.

The peninsula of India is commonly called Deccan, whose northern boundary was marked by the rivers Nerbudda and Mahanudda. In those days this vast track of territory was divided among the five different nations. Those were the Murhuttas of Marhutt, the Tellingas of Telling, the Cunnuras of Cunnur, the Goonds of Goondwarm and the Tamoolas of Tamool. Each of these nations were probably a separate monarch.

People of every nation were different from each other on the ground of their manners, habits, dresses and language. People were divided from its neighbour either by broad river, thick forest, or stupendous mountain.

Deccan land have been very important in life of Banjara community because they migrated in large numbers from North India to Deccan, their traditional occupation became prosperous in Deccan, many occupational groups joined them in Deccan, they centralized in Deccan in a large number and finally began to settle down in Deccan, when they lost their traditional occupation. The Deccan land was most suited for their traditional occupation. The following were the reasons behind it. Firstly, political stability, no war between different political powers, took place late in Deccan then in other parts of country. In the period of political instability there was continuous demand for their services in war time. History shows that Banjaras were engaged in supplying foodgrains to army camp up to 1803 in large numbers[5]. Secondly, Deccan was divided into five different provinces. The uncertainty of rains frequently was responsible for the famines. Under such condition there was always need of Banjaras for internal trade. Thirdly, the wild tribes of Deccan were far away from Market mechanism. There was scope to bring wild products of wild tribes in market and develop new wants and tests among such wild tribes. Banjaras had capacity to enter in their territories and bring them into market mechanism[6]. Finally, there was no such class in Deccan, who could do such type of work in large scale. It means the field was open for Banjaras for their traditional occupation. Because of this reason there was monopoly of Banjaras in trade and carrying occupation in Deccan before modern means of transport like railway, came into existence.

COLLAPSE OF TRADITIONAL OCCUPATION

Political instability (wars between different political powers), absence of road development, absence of modern means of transport like railway and absence of competition were basic support of traditional occupation of this community. But in first half of nineteenth century they almost lost these supporting factors. This change badly affected their traditional occupation and took away their bread and means of livelihood.

The first attack, on their traditional occupation began to start when political stability took place in our country. In 18th century and before there was continuous war between different political powers and therefore there was continuous demand of services of Banjaras. But in first half of nineteenth century British succeeded to establish their political power in most part of the country. Naturally, this political stability reduced the services of Banjaras at war time.

The third and very important attack, which badly affected their long distance trading was development of railway in our country.

Development of railway is the factor which is not only badly affected Banjaras traditional occupation, but totally changed carrying and trade pattern in our country. Following some records show their importance and effect of railway on them. In 1836, about 14,000 tonnes of cotton went through the Nasik district from Berar region to Bombay and in nine years following, the average quantity was about 15,220 tonnes. This cotton was carried chiefly by pack of the bullocks. It was estimated that not less than 1,80,000 bullocks were employed. In 1862 almost the whole of Berar and Khandesh cotton crop passed to the sea by rail[7].

The fourth attack was competition began to start in trade and carrying occupation. In former time, for trade and carrying purpose traveling in forests and hilly tracts with bullocks under unhealthy climate was very essential. So, it was a very hard and risky occupation. Apart from this there was very less possibility of profit. Due to this reason, there was no competition in this occupation. But, when modern means of transport came into existence, the occupation became more safer than before and possibility of profit also increased. Because of these reasons many local groups and outsiders entered in this occupation.

As stated earlier, political instability, absence of road development, absence of modern means of transport and absence of competition were basic supporting factors. But when they lost these factors, naturally, they lost their traditional occupation, without adjusting with new mode of trade they turn to various occupation and finally on agriculture specially in hilly tracts.

WHY BANJARAS TURNED TO AGRICULTURE?

Today, agriculture is main occupation of this community, especially in Maharashtra. It is important to know in this study, that a community who was dominated in trade and carrying occupation in many centuries. There was almost monopoly in this occupation of this community. When this occupation took new mode, without adjusting this new mode of occupation, why this community turned to agriculture for their livelihood? Today, Banjaras are settled down specially in hilly tract, settlement in hilly tract and dependence on agriculture two factors are related to each other.

In middle of nineteenth century they lost their traditional occupation. The period, between middle of the nineteenth century to middle of twentieth century was very crucial for this community, when they were engaged in their traditional occupation, they never though about alternate to this occupation. This occupation was suitable for their way of life and they totally adjusted to this occupation. When they suddenly lost this occupation they became routeless and turned to various occupations for time being. These means of livelihood were not sufficient and resulted into more poverty and unemployment. Naturally, this poverty and unemployment turned them to crime. At the same time British Government declared them as criminal tribe under Criminal Tribes Act 1871. Up to 1952 they were under dark shadow of this act. About one hundred years they were under this adverse condition and finally settled down in hilly tracts and depended directly and indirectly on agriculture. Following are the reasons why a trading community settled down in hilly tracts and turned to agriculture.

CATTLE PROPERTY

Cattle property always played an important role in economic life of this community. The history of this community shows that when they migrated with Mughal army from North to South the Sarang and Jhungi (leader of this community) had 1,80,000 bullocks with them and at the same time Bhagwandas (another leader) had 52,000 bullocks[8]. Another record shows that British first time employed this community in 1791-92 in the war which took place against Tippu Sultan. Banjaras joined this war, as grain suppliers, with their 60,000 bullocks. Again they joined in 1798 with their 25,000 bullocks, when battle was going in Srirangapatnam (capital of Tippu). Thus the traditional occupation was linked with their cattle property[9].

In history, Banjaras are known by their traditional occupation, carrying goods on pack-bullocks. Today, the main occupation of this community is agriculture. But historical records show that this community tried (gone through) for various occupations (professions), that are food gathering, gazier, as guards to travelers and goods, carrying goods on pack-bullocks, supplying foodgrains to army camp, tenants and finally as farmer. It is important to note here that in every phase of their economic life cattle played an important role. When they lost their traditional occupation, they turned to agriculture and settled down in hilly tracts with hoping better grass and water facilities for their cattle.

LITTLE EXPERIENCE OF AGRICULTURE

After collapse of their traditional occupation, they had turned to agriculture. It does not mean that they had selected a totally new occupation. In former time, because of nature of their traditional occupation, they were always nomadic. In this process they always halted for a short period (some time one or two years). In this time they always got land on rent for season. In this land they always used to grow (sesame), an oil seed, because this oilseed can be raised without much labour or skill and could grow on any poor strip of land[10]. From the selection of crop in nomadic time a few economic inferences could be drawn. Firstly they always had halted at poor land away from settled population which was suitable for their cattle (water and grass facilities). Secondly, they were habitual to temporarily settle on poor lands. Thirdly, because of this they started to such crop which was suitable to grow in poor lands. Because of this habit and little experience they turned to agriculture and settled down in hilly tracts.

FALL OF PROFIT

In former time the nature of trade and carrying occupation was very hard and risky. So there was no alternative to Banjaras service. So Banjaras were getting sufficient to fulfill their limited needs. But when new means of transport came into existence it created competition in this occupation. Because of this competition, rate of profit began to decline, especially trade on the back of bullocks became an outdated business of very marginal profit. When their chiefs found that tillage (an oilseed), this oilseed can be raised without much labour or skill and could grow on any poor strip of land, paid better than grain carrying they began to settle as landholders.

CRIMINAL TRIBES ACT

India is probably, the only country in the world that recognizes crime as an inherited trail and number of men, women and even children are registered and stigmatized as criminal from birth. In 1871, British passed the Criminal Tribes Act. It notified about 150 tribes in India as criminal, given the police wide power to arrest them and monitor their movement[11]. In 1925, the total population of notified groups under this Act was estimated at about 40 lakh[12]. The effect of this law was simple; just being born in one of those 150 tribes make you a criminal. Like other tribes Banjara community was living under dark shadow of this Act for about 80 years.

This Act lays down four distinct progression stages of treatment: (i) notification of criminal tribes, (ii) registration of members of notified tribes, (iii) restriction on movement of notified tribes, and (iv) settlement of notified tribes. Settlement of criminal tribes was final and important stage of this act. But without giving much attention to this stage, the British Government mainly concentrated on restriction of criminal tribes. The most inhuman in this act was every registered member of notified criminal tribes had to report himself at fixed interval and the power of taking 'Hazari' was given in the hands of village headman[13]

As stated above, restriction on movement of criminal tribes was one of the important part of this Act. Banjaras being declared as a criminal community, the restriction on their movement came directly. When they lost their traditional occupation they were in search of proper occupation and place for settlement. But at the same time this Act was imposed on them and they were indirectly forced to settle down where they were.

Generally, Banjaras have settled down in hilly tracts on account of oxen with them. But this is not the only reason behind their settlement pattern. The Criminal Tribes Act is also responsible for this settlement pattern. Because of stigma of this Act, restriction on movement and 'Hazari' system they always avoided police and settled people. This is also one of the reasons why they settled down near hilly tracts, away from settled people, and finally turned to agriculture.

BANJARA COMMUNITY IN MAHARASHTRA

On 1st May 1960, the new Maharashtra State came on the map of India. This Maharashtra State is divided among following different regions: Marathwada, Vidarbha, Western Maharashtra, Konkan and Khandesh. The most part of Maharashtra were formerly in those states where Banjaras were mostly located, for example, Marathwada region was formerly in Nizam's state (Hyderabad state), Vidarbha was in Berar and Khandesh was in Bombay state. Now a days, Banjaras of Maharashtra are mainly located in Marathwada, Vidarbha and Khandesh. Because of this, Maharashtra is proper state for the study of Banjara community.

POPULATION OF BANJARA COMMUNITY

This study is related to Banjara community of Maharashtra. But before we see the population structure of Banjaras in Maharashtra, if we see the population structure of this community at all India level, we can understand more about this community. Above we have made a statement that Maharashtra is important state for study of this community, because the Maharashtra is divided among five regions, and most part of the Maharashtra were formerly in those states where Banjaras were mostly located. For justifying the above statement we have to see the population of Banjara community at all India level.

Table 1 shows that in the years 1911, 1921 and 1931 the Banjara population was divided among following states or agencies: Bombay, Central Provinces and Berar, Mysore and Rajputta agency. The table shows that in the year 1911 the total population of Banjara community was 8,37,155. Out of this total population, the highest population (52 per cent) was in Hyderabad State. After Hyderabad state, C.P. and Berar was on second number, it was 20.25 per cent of total population. On the third place there was Bombay state (17.15 per cent). On the fourth place was Mysore state (6.1 per cent) and lastly Rajputana agency (4.49 per cent).

The table shows another fact that although the total population of Banjara decreased in 1931 as compared to population of 1911, from 8,37,155 to 7,68,988, the percentage of Banjara population in every state compared to total population remained more or less the same.

Table 1
Population of Banjara Community in India

State	*Population in*		
	1911	*1921*	*1931*
Bombay (including W.E.S. agency)	143586 (17.15)	109025 (18.78)	143317 (18.64)
Central Provinces and Berar	169505 (20.25)	122973 (21.18)	155353 (20.20)
Hyderabad State	435322 (52.00)	263844 (45.45)	371402 (48.30)
Mysore State	51168 (6.11)	53300 (9.18)	63368 (8.24)
Rajputana agency	37574 (4.49)	31436 (5.41)	35548 (4.62)
Total	837155 (100)	580578 (100)	768988 (100)

Source: Census of India, 1931, Vol. I, India, Part I, Chapter XII, pp. 462-468, published by Government of India.

The table 1 also shows that although the population of Banjara community decreased in 1921 as compared to population of 1911 and again it increased in 1931 as compared to population in 1921, the percentage of Banjara population in Hyderabad state was always higher than that in other states. Table 1 also shows that there is slight change in total Banjara population and percentage of Banjara population of every state to total population in every census year. But the ranking of states according to percentage of Banjara population has not changed. During all the census years Hyderabad state ranked first, C.P. and Berar ranked second, Bombay ranked third, Mysore ranked fourth and Rajputana agency is last.

In the historical part of this study, we have seen that North India, Rajputana, is original home of Banjara community, from where they migrated to different parts of the country. But it is seen from the table 1 that in original home of Banjara community the population was very small compared to other states. For example in 1911 only 4.49 per cent Banjara population was in their original home, in 1921 it was 5.41 per cent and in the year 1931 it was 4.62 per cent. It shows how Banjaras have migrated in large numbers from North to South India. We have also seen in historical part of this study that they systematically migrated with Mughal power. And up to 1948 there was power of Nizam (Mughal) in Hyderabad state. It shows that Banjaras of Deccan were mainly centralized in those areas where Mughals were in power, because in Nizam's state (Hyderabad) about 50 per cent population of Banjaras was centralized in 1931.

BANJARA POPULATION OF MAHARASHTRA

Before independence individual caste and tribe base was considered for the census of population. But after independence, as a policy matter, there is no census of population according to individual caste and tribe. So, Banjara as a caste or tribe there is nothing separate in census report after independence. But there is one alternative to solve this problem. Although there is no census according to individual caste and tribe there is census of population according to mother tongue. With the help of mother tongue criteria we can find out the total population of Banjaras in Maharashtra. For Banjara population we take the support of mother tongue and census report of Maharashtra, 1961. Following are the reasons for considering these two factors:

First reason is, Banjara dialect is unique one. It is mixture of Rajasthani and Gujarathi languages. This dialect is known as Banjari/ Lambad. The important thing about this dialect is that all over India this dialect is used by Banjara people among themselves. In the time of process many customs and traditions of this community becoming out of picture but they still preserved their dialect. Second reason is census report of 1961 is proper year to census the Banjara population of Maharashtra. If we take other years like 1971, 1981, 1911, it will not give proper picture of the total population of this community according to mother tongue. The reason is that from last fifty years the Banjaras of Maharashtra have very much mixed with other people and adjusted themselves. Naturally the influence of Marathi language is very much on them. And most of them, in recent years, say that Marathi is their mother tongue, which was not happening fifty years ago. So for the population of Banjara community in Maharashtra if we take mother tongue criteria and the year 1961, it will give proper picture of Banjara population in Maharashtra.

We have already seen the justification about why Mother tongue of this community and year 1961 considered for the study of population of Banjara community in Maharashtra. The thirteen districts shown in table 2 and 3 were having the larger Banjara population according to 1961 population census. The Banjara population according to Mother tongue was not recorded for the remaining districts of Maharashtra. So for our analysis we considered only those districts in which Banjara population is shown according to Mother tongue Banjari/Lambadi. The districts which are shown in Tables 2 and 3 can be divided according to regions for the analysis of population. Aurangabad, Osmanabad, Parbhani, Nanded and Beed districts belong to Marathwada region. Yeotmal, Wardha, Amravati, Akola and Buldhana districts belong to Vidarbha region. Jalgaon and Dhulia belong to Khandesh region. Only Solapur district belongs to Western Maharashtra region.

POPULATION PICTURE AT REGIONAL LEVEL

Table 2 shows the percentage of Banjara population to total population of the districts in the year 1961. As stated earlier, first five districts belong to Marathwada region. The total population of Marathwada region was 62,97,373. Out of this total population, 1,52,270 were Banjaras. It means that in Marathwada the percentage of Banjara population was 2.42. In Vidarbha region (second five districts) the total population was 52,14,579. Out of this, 1,61,060 were Banjaras. It means in Vidarbha region, the Banjara population was 3.09 per cent of total population. In Khandesh, the percentage of Banjara population to total population was 1.06. In Western Maharashtra Banjara population was shown in only Solapur district. Out of total population of Solapur, 18,60,119, the Banjara population was 7354. It means in Solapur district, the Banjara population was only 0.40 per cent.

Table 2
Banjara Population (According to Mother Tongue) in Selected Districts of Maharashtra in 1961

District	*Total population of the district*	*Banjara population*	*Percentage of Banjara population to total population*
Aurangabad	1532341	35893	2.34
Osmanabad	1477656	13635	0.92
Parbhani	1206236	37148	3.08
Nanded	1079674	49179	4.55
Beed	1001466	16415	1.64
Yeotmal	1098470	103318	9.40
Wardha	634277	2022	0.32
Amravati	1232780	3116	0.26
Akola	1189354	47711	4.01
Buldhana	1059698	4893	0.46
Jalgaon	1765047	25734	1.46
Dhulia	1351236	7229	0.53
Solapur	1860119	7354	0.40
Total	1,64,88,354	3,53,647	2.14

Source: Census of India 1961, District Census Hand books, Government of Maharashtra.

Now lest us see the percentage of Banjara population to total Banjara population at regional level. Table 3 shows that the total population of Banjaras in thirteen districts was 3,53,647. In Marathwada region, the Banjara population was 1,52,270 which was 43.06 per cent of total Banjara population in the state. In Vidarbha region Banjara population was 1,61,060 which was 45.54 per cent of total Banjara population in the state. In Khandesh region the Banjara population was 32,963 which was 9.32 per cent of total Banjara population in the state. In Western Maharashtra (Solapur District) the

Banjara population was 7,354 which was 2.08 per cent of total Banjara population of the state.

Table 3
Districtwise Percentage Distribution of Banjara Population (According to Mother Tongue) in Maharashtra in 1961

District	*Total population of Banjaras*	*Percentage of Banjara population to total Banjara population*
Aurangabad	35893	10.15
Osmanabad	13635	3.86
Parbhani	37148	10.15
Nanded	49179	13.91
Beed	16415	4.64
Yeotmal	103318	29.22
Wardha	2022	0.57
Amravati	3116	0.88
Akola	47711	13.49
Buldhana	4893	1.38
Jalgaon	25734	7.28
Dhulia	7229	2.04
Solapur	7354	2.08
Total	3,53,647	100.00

Source: Census of India 1961, District Census Hand books, Government of Maharashtra.

POPULATION PICTURE OF DISTRICT LEVEL

We have seen above the pattern of Banjara population at regional level. Now with the help of Tables 2 and 3 we can see the percentage of Banjara population to total population and percentage of Banjara population the total Banjara population in 1961 in thirteen districts of Maharashtra.

Table 2 shows that the Banjara population in thirteen districts of Maharashtra forms 2.14 per cent of total population of Maharashtra. Yeotmal, Akola, Nanded, Parbhani and Aurangabad are above the average percentage. Remaining districts are below the average. Table 2 also shows that the percentage of Banjara population to total population is highest in Yeotmal (9.40), the second is Nanded (4.55), the third is Akola (4.01), the fourth is Parbhani (8.08), and Aurangabad is on fifth place (2.34). Table 2 also shows that the percentage of Banjara population in Osmanabad, Wardha, Amravati, Buldhana, Dhulia and in Solapur is less than one per cent.

Table 3 shows the districtwise percentage of Banjara population in a district to total Banjara population, Yeotmal is at the top (29.22 per cent), the second is Nanded (13.91 per cent), the third is Akola (13.49 per cent), the fourth is Parbhani (10.15 per cent). The lowest population of Banjara is in Wardha and Amravati districts (less than one per cent).

PHASES OF ECONOMIC EMERGENCE OF BANJARAS IN MAHARASHTRA

On historical records Banjara community is known as carriers of goods on pack-bullocks. Till this day this community is mostly settled down in Maharashtra and mostly depend directly and indirectly on agriculture. The historical record of this community shows that it tried (gone through) for various occupations (professions) before settling down as farmers. But it looks from their history that always decisive factor of their occupation was the ownership of cattle. Following are the various phases from which they had gone through before settle down as farmers.

First Phase [Food Gathering or Pastoral Nomad]

It is true that Banjaras used to carry the goods on oxen-backs, from long distances. They were engaged for many centuries in this occupation. This is the reason that this occupation is known as their traditional occupation. But this was not first phase of their life. Before entering in this phase they were wandering with their cattle in forest. Because of this they are known as Charan Banjaras. Kutch was mainly their region, because many customs of Charan Banjaras of Maharashtra are linked with Kachchela Charan of Gujarat. Kachchela means from Kutch. In this phase forest product and livestock economy was initial occupation of this division. Thus wandering in forest with cattle in search of food was the beginning phase of economic emergence of these people. In this phase they were nomadic. We can call them in this phase as pastoral nomads or food gathering nomads[14].

Second Phase [Trade Nomads–Nomads with Limited Zone]

In the first phase, while wandering with cattle, some feature developed among them which supported them to enter in second phase. While wandering they had cattle property and they became familiar with geographical routes. They were not only familiar to geographical routes but developed the capacity to fight in adverse condition. From the very beginning, honesty in dealing with others was among them. In very former time travelers and goods carriers were taking the support of Banjara tandas for transport of their goods from one place to another from safety point of view. And naturally, afterwards they started the work of guards of travelers and goods. The experience as guards brought them near to their traditional occupation. Later, having cattle and honesty in dealing they themselves started the carrying occupation. Here again we see that decisive factor for their entry in this second phase was cattle ownership[15].

Economic activities started to increase in north India and naturally it promoted trade and commerce activities. Under such conditions the bullocks became an important means of transport of goods. Naturally, having bullock property, Banjaras got their new occupation. But it was a fact that because of slow growth of economic activities their utility was limited with north India, specially Gujarat and Rajasthan, in other words, north-western zone. This phase was of many centuries. Being many centuries they were in this zone. Naturally, there is greater influence of Rajasthan and Gujarat culture on the culture of Banjaras. Their dress, the colour of dress, dialect, customs, traditions or in other words, total culture was greatly influenced by the culture of these two regions[16].

Third Phase [Trade Nomads with No Limitation]

This phase is very important for Banjara community. Banjaras entered in this phase because of mainly three reasons: (1) Banjaras had experience of carrying goods on pack-bullocks in north and had much bullock property, (2) Mughals wanted to politically step down in south India and for this purpose they were in search of such class who could provide them foodgrains for army for at the time of war, and (3) there was no such specialist class in South India who could do such type of occupation. In this phase Banjaras were for about two hundred years. Because of the role they played in this phase they were known in history, not only this that they came on historical record. In this phase they migrated from north to south and became trade nomads of no limitation. Following are the main features of this phase:

1. **No Attachment to Political Power:** Migration from north to south was the beginning of this phase. This was the time Mughals wanted to step down in South India. Banjaras migrated with them as carriers. In this phase not only Mughals but Marathas, Rajputs, Peshwas and lastly British too utilized their services. In those days, no other community came in contact with various political powers like Banjaras[17].
2. **Occupational Approach:** It is a fact that they came in contact with different political powers; but it is also true that they were not attached politically to any power. Because of this, about every political power utilized their services and as a result their carrying occupation increased. For this occupation they reached upto Madras in this phase. They gave more importance to their occupational interest than political.
3. **Occupational Peak Point:** This is the phase where their traditional occupation was on peak point, because on one side due to political instability there was continuous demand for their services and on other side the field was open for trade in central and south India. This is the reason that there was monopoly of these people in trade and carrying occupation.

4. **Expansion of Tandas:** This phase was peak point of their occupation. So, naturally, the carvans (tandas) had capacity to provide services to other occupation groups. Because of this, many occupation groups joined Banjaras and carvans began to take big shape.
5. **Tribal Organization:** In this phase their tribal organization was very strong. The whole community was under the control of Chief Naik. In this phase they formed their little government with social rules and regulation. The whole community was governed by these social rules and regulation and the word of supreme (Naik) was final to any disputes. Panchayat system was one of the features of their little government[18].
6. **Culture and Religion:** They strictly preserved their culture in this phase. They never came into close contact with other people and this is the reason that there was no influence of other culture on the culture of Banjaras. As foodgrain suppliers to army camp, they came in more contact with Mughal power than any one. But it is important to note here that they always kept themselves away from Muslim religion.

Fourth Phase [Tenants]

Political instability and non-availability of modern means of transport were the basic reasons which explain why the traditional occupation of Banjaras reached to a peak point in the third phase. But these two factors began to change almost at the same time. On one side most part of India came under power of British Government and political stability began to start. British started to exploit this stability for their commercial purpose. Development of railway was result of it. This change badly affected traditional occupation of Banjaras. In very short time they lost their traditional occupation in which they were engaged from many centuries. This change pushed them into fourth phase as tenant.

After collapse of their traditional occupation they started working as tenants of the landlords. The basic reason for this was again the cattle ownership. The skill of cultivation of land was quickly acquired by them. It is well known that a migrating community when try to settle down works more efficiently to improve their economic status. Banjaras were no exception to this universal value. Until the third stage the concept of creation of wealth was not properly developed among them; because of their nomadic life. But when they started to settle down this concept started to develop among them.

This was the phase that Banjaras first time became rootless. Because of sudden collapse of their traditional occupation they turned to crime to some extent and stigma of criminal tribe was stamped on them by British government which they cannot wash out totally till today. This phase was in between nomadic and settled life in which they were struggling for alternate occupation to their traditional occupation. In this phase they totally depended on settled people for their livelihood. Every small tanda found their means of livelihood by their own way. In this process they began to settle down in hilly tracts. When they lost their traditional occupation they also lost the control of their supreme on community and became unorganized. This was the phase when they first time came into contact with settled people and their culture. This close relation with other people and culture began to start effect on culture of Banjaras. Apart from above unfavourable conditions, with their hardworking nature and capacity to fight against adverse conditions they came out of this phase and entered in the fifth phase.

Fifth Phase [Farmer]

Today Banjaras of Maharashtra have settled down as a stable community in rural Maharashtra. Most of them are cultivators and agricultural labourers. It is noteworthy that the community that was entirely landless about a century ago; but now most of the Banjaras are land owners. This happened because of their approach of fighting in adverse conditions, their hardworking nature, their transport capacity and ownership of bullocks. After having lost their traditional occupation they gave more importance to creation of wealth and created wealth in the form of land. But it is also the fact that their holdings are small and land is not so fertile. It happened because of their settlement in hilly tracts. This change is also closely related to their cattle.

Recently, with the development of sugar factories, one important occupational change took place in Banjaras of Maharashtra. No doubt most of the Banjaras have gained land through their hard working. But their holdings are small and a good number of them are agricultural labours. Demand for labour created by sugar factories provided an opportunity for the Banjaras to use their migratory nature and same groups have started migrating to the sugar factories in search of work. This is the seasonal occupation that they started to support their economical life. Here also the change of occupation is related to ownership of cattle and capacity of transport.

Because of changing phases of Banjaras of Maharashtra sometimes they are called as "Bha-ru-pi" Banjaras. The meaning of this word is many modes of life.

A study to look into the detailed aspects of the economics of Banjara community in Maharashtra was conducted. This study was based mainly on primary data, which was collected through canvassing a questionnaire among the important persons of sample tandas. As per the population census of 1961 there are thirteen districts in Maharashtra State where the Banjara community is largely concentrated. These districts are: Beed, Solapur, Osmanabad, Latur, Jalna, Aurangabad, Parbhani, Nanded, Buldhana, Jalgaon, Yeotmal, Nasik and Akola. Multi-stage stratified random sampling technique was used to select the ultimate sampling unit i.e. Banjara tanda. This study is based on primary data collected from 118 tandas of thirteen districts of Maharashtra where Banjaras are mainly located. The unit of analysis is at the level of tanda and not the household. While selecting the Banjara tandas for the survey work due attention was paid towards their population size, distance from taluka place, distance from forest and year of settlement.

LAND HOLDING

The districtwise distribution of families of sample tandas according to landholding is presented in Table 4. It is evident form this table that a maximum number of families (3151 families or 25.42 per cent) were holding land up to 5 acres. This proportion was even higher than the figure of 25.42 per cent in the case of sample tandas in Beed (30.01 per cent), Parbhani (34.94 per cent), Aurangabad (42.39 per cent), Jalna (34.69 per cent), Nanded (24.95 per cent) and Nasik (38.60 per cent) districts. 17.77 per cent of total families of sample Banjara tandas in all the thirteen districts were having landholding between 5 and 10 acres. This percentage was even higher than this in the case of sample tandas in Yeotmal (19.76 per cent), Akola (19.89 per cent), Aurangabad (19.39 per cent), Parbhani (23.44 per cent), Latur (27.16 per cent), Osmanabad (33.44 per cent) and Beed (26.42 per cent) districts. 15.03 per cent of the total families were having land between 10-15 acres in the sample Banjara tandas of Maharashtra. The figure was higher than 15.03 per cent in the sample Banjara tandas of Beed (24.62 per cent), Aurangabad (15.71 per cent), Nanded (15.25 per cent), Buldhana (23.16 per cent), Nasik (15.80 per cent), Akola (18.16 per cent) and marginally higher in Yeotmal (15.16 per cent) district. 10.15 per cent families were holding land between 15-20 acres and 6.82 per cent families were holding land above 25 acres in sample Banjara tandas of Maharashtra. The percentage of families holding land between 15-25 acres was relatively higher in the sample Banjara tandas of Jalna (16.60), Buldhana, (9.83), Akola (17.47) and Yeotmal (8.12) district.

It is important to note that about one-fourth (24.81 per cent) of the total families in sample Banjara tandas in thirteen selected districts of Maharashtra were not having any land of their own. It means one-fourth of the families of sample tandas were landless. The percentage of landless families was even higher than this (24.81 per cent) in the Banjara tandas of Solapur (51.41 per cent), Osmanabad (35.31 per cent),

Latur (29.14 per cent), Nasik (26.68 per cent), Jalgaon (51.54 per cent) and Yeotmal (28.48 per cent) district. From the above analysis it seems that one-fourth of the families were landless, another one-fourth of the families were small farmers, the remaining 32 per cent families were large farmers having land above 10 acres. It looks from the above figure that about 75 per cent of Banjara families have agriculture land. It shows that a community, which was about landless about 100 years ago, now majority of them are land owners. The main reason behind it is that after losing their traditional occupation they gave more importance to create permanent income source in the form of land, which was not possible in their nomadic life and because of this they turned to crime after loss of their traditional occupation.

Table 4
Land holding Pattern of Banjara Families in Sample Tandas of Maharashtra, (Acres)

Districts	*Total No. of families*	*Landless*	*Up to 5*	*5 to 10*	*10 to 15*	*15 to 25*	*Above 25*
Beed	723	36 (4.98)	217 (30.01)	191 (26.42)	178 (24.62)	70 (9.68)	31 (4.29)
Solapur	815	419 (51.41)	144 (17.67)	63 (7.73)	65 (7.98)	71 (8.71)	53 (6.50)
Osmanabad	320	113 (35.31)	14 (4.38)	107 (33.44)	24 (7.50)	55 (17.19)	7 (2.18)
Latur	405	118 (29.14)	69 (17.04)	110 (27.16)	58 (14.32)	33 (8.15)	17 (4.19)
Parbhani	1182	222 (18.78)	413 (34.94)	277 (23.44)	145 (12.27)	79 (6.68)	46 (3.89)
Aurangabad	1248	144 (11.54)	529 (42.39)	242 (19.39)	196 (15.71)	95 (7.61)	42 (3.36)
Jalna	542	60 (11.07)	188 (34.69)	38 (7.01)	75 (13.84)	91 (16.79)	90 (16.60)
Nanded	1515	233 (15.38)	378 (24.95)	276 (18.22)	231 (15.25)	267 (17.62)	130 (8.58)
Buldhana	885	104 (11.75)	149 (16.84)	130 (14.69)	205 (23.16)	210 (23.73)	87 (9.83)
Nasik	386	103 (26.68)	149 (38.60)	47 (12.18)	61 (15.80)	25 (6.48)	1 (0.26)
Akola	1041	191 (18.35)	201 (19.31)	207 (19.89)	189 (18.16)	71 (6.82)	182 (17.47)
Jalgaon	1661	856 (51.54)	371 (22.34)	184 (11.08)	182 (10.96)	44 (2.65)	24 (1.43)
Yeotmal	1675	477 (28.48)	329 (18.64)	331 (19.76)	254 (15.16)	148 (8.84)	136 (8.12)
Overall	12398	3076 (24.81)	3151 (25.42)	2203 (17.77)	1863 (15.03)	1259 (10.15)	846 (6.82)

Note: Figures in brackets indicate percentage to total.
Source: Field survey

DIVISION OF QUALITY LAND OWNED BY DWELLERS

The Banjara tandas are located mostly on hill sides or at the foot of hills. As a result, the agricultural land at the command of Banjara is of low quality and less fertile. The data about the quality of land with Banjaras in Maharashtra presented in Table 5 also indicates the same thing. It could be seen from this table that only 20 per cent of land owned by dwellers was fertile. The reason behind it is that because of settling down in hilly tracts, naturally they got less fertile land. Their poverty also not allowed them to purchase fertile land. About 46 per cent of total land owned by them was not arable. 36 per cent of land owned by Banjaras of sample tandas was suitable to grow one crop in a year. The above figures reveal that the Banjara tandas in Maharashtra have very little fertile land and this affects the crop system and crop intensity in the agricultural land of Banjara tandas. The dwellers in Jalgaon district have the maximum percentage (32.08 per cent) of fertile land followed by the Banjaras in Jalna (30.70 per cent), Aurangabad (26.15 per cent), Yeotmal (24.58 per cent), Akola (20.50 per cent), Buldhana (20 per cent) and Nanded (20 per cent) districts. The Banjaras of sample tandas in Osmanabad (5 per cent), Solapur (6.67 per cent), Beed (9 per cent) and Nasik (6.25 per cent) districts have relatively very low proportion of fertile land at their command. Naturally, the economic condition of the Banjaras residing in the tandas of these districts might be relatively weaker as compared to those residing in tandas of other sample districts with higher proportion of fertile land. The percentage of less fertile land owned by the dwellers was relatively higher in the case of Banjara tandas in Buldhana (47.14 per cent), Parbhani (40.90 per cent), Aurangabad (40 per cent), Latur (38.33 per cent), Yeotmal (38.33 per cent) and Osmanabad (36.25 per cent) districts as compared to the figure of 34.20 per cent for all the sample districts. The district of Nasik (12.50 per cent), Solapur (23.34 per cent), Beed (27 per cent) and Nanded (30.60 per cent) were below the average figure of 34.21 per cent. The land of Banjara tandas in Solapur (70 per cent), Beed (64 per cent), Nasik (81.25 per cent), Osmanabad (58 per cent), and Nanded (49.40 per cent) districts was mostly of low quality and not arable. The economy of the Banjara tandas located in these districts must be relatively weaker as compared to that of Banjara tandas in the districts having relatively higher proportion of land suitable for growing one or two crops in a year.

PERENNIAL AND SEASONAL IRRIGATED AREA

The districtwise data about perennial and seasonal irrigated area in sample Banjara tandas of Maharashtra is presented in Table 6. It is evident from the table that out of the total agricultural land of the Banjara tandas (1,02,066 acres), only 4414 acres of land (4.32 per cent) was under perennial irrigation and 6700 acres (6.56 per cent) were under seasonal irrigation. It means that about 10.88 per cent of the agricultural land of sample Banjara tandas was either perennial or seasonally irrigated during the survey year. And the remaining 89.12 per cent agricultural land was dry. The Banjara tandas in the districts of Osmanabad (6.31 per cent), Aurangabad (5.91 per cent), Jalna (8.48 per cent), Nasik (5.13 per cent), Nanded (4.81 per cent) and Jalgaon (4.96 per cent) were having relatively higher proportion of area under perennial irrigation than the figure at the all districts level (4.32 per cent). The districts of Beed, Solapur, Latur, Parbhani and Yeotmal were well below the all district level figure. Buldhana and Akola districts were very close to the all district level picture. It is to be noted that the agricultural land under perennial irrigation of tandas in Solapur district was as low as 1.57 per cent and it was as high as 8.48 per cent in Jalna district.

Table 5
Percentage Distribution of Agricultural Land of Sample Tandas in Maharashtra

Districts	*Fertile land*	*Less fertile land*	*Low fertile land*
Beed	9.00	27.00	64.00
Solapur	6.66	23.34	70.00
Osmanabad	5.00	36.25	58.75
Latur	17.50	38.33	44.17
Parbhani	16.80	40.90	42.30
Aurangabad	26.15	40.00	33.85
Jalna	30.70	35.00	34.30
Nanded	20.00	30.60	49.40
Buldhana	20.00	47.14	32.86
Nasik	6.25	12.50	81.25
Akola	20.50	39.00	40.50
Jalgaon	32.08	26.25	41.67
Yeotmal	24.58	38.33	37.09
Overall	20	34.2	45.8

Source: Field survey.

Regarding the percentage of land under seasonal irrigation, it is seen from the table that sample Banjara tandas of Beed (6.88 per cent), Osmanabad (18.73 per cent), Parbhani (7.73 per cent), Aurangabad (11.94 per cent), Akola (8.73 per cent) and Jalgaon (8.59 per cent) districts were having the proportion of seasonally irrigated land higher than the all district level figure of 6.56 per cent. The sample Banjara tandas of Solapur, Latur, Jalna, Nanded, Buldhana, Nasik and Yeotmal districts were well below the all-district level figure.

Considering the perennial and seasonal irrigated areas together, it seems that the Banjara tandas of Osmanabad, Aurangabad, Jalna,

Akola and Jalgaon districts were well above the all-district level figure, and Banjara tandas of Beed and Parbhani districts were almost around this figure. The Banjara tandas of Solapur, Latur, Nanded, Buldhana, Nasik and Yeotmal districts were well below the all-district level figure and hence irrigationally backward.

Table 6
Perennial and Seasonal Area Under Irrigation in Sample Tandas of Maharashtra

Districts	*Total land of sample tandas*	*Perennially irrigated land*	*Seasonally irrigated land*
Beed	6763	335 (3.95)	465 (6.88)
Solapur	4791	75 (1.57)	280 (5.84)
Osmanabad	2536	160 (6.31)	475 (18.73)
Latur	3011	72 (2.39)	110 (3.65)
Parbhani	8308	272 (3.27)	642 (7.73)
Aurangabad	9227	545 (5.91)	1102 (11.94)
Jalna	6428	545 (8.48)	315 (4.90)
Nanded	15600	750 (4.81)	790 (5.06)
Buldhana	10962	465 (4.24)	465 (4.24)
Nasik	2146	110 (5.13)	96 (4.47)
Akola	11566	480 (4.15)	1010 (8.73)
Jalgaon	6751	335 (4.96)	580 (8.59)
Yeotmal	13977	270 (1.93)	370 (2.65)
Overall	102066	4414 (4.32)	6700 (6.56)

Note: Figures in brackets indicate percentage to total.
Source: Field survey.

CROP PATTERN

The districtwise data about cropping pattern in sample Banjara tandas of Maharashtra are shown in Table 7. The percentage area under different crops in sample tandas is shown in this table. A look at the table allows us to state that Kharif jowar and cotton were the main crops grown by the families of sample Banjara tandas in Maharashtra during the survey year. As much as one-third (33.09 per cent) of the total cropped area of sample tandas was under kharif jowar. This percentage of area under kharif jowar was even higher than this figure in the Banjara tandas belonging to Solapur (51.70 per cent), Latur (46.60 per cent), Nanded (35.30 per cent), Jalna (35.00 per cent) and Parbhani (37.70 per cent) districts. It seems that the Banjara tandas belonging to Marathwada region and adjoining Solapur district were growing kharif jowar as their main crop. The farmers residing in Banjara tandas of these districts have low quality lands and have very less irrigation facilities. As a result, they have to put a larger part of their agricultural land to cultivation of kharif jowar crop.

The second important crop grown by the farmers of sample Banjara tandas in Maharashtra is cotton. 25.62 per cent of the total cropped area of sample Banjara tandas was under cotton crop. The percentage area under cotton was even higher than this in the sample Banjara tandas of Yeotmal (66.20 per cent), Akola (47.0 per cent), Nanded (43.80 per cent), Parbhani (34.50 per cent), Buldhana (34.30 per cent) and Jalna (32.10 per cent) districts of Maharashtra state. These districts of Marathwada and Vidarbha regions are also known as the cotton producing districts. So, the farmers of sample Banjara tandas from these districts also found to have put larger area of their cultivated land to cotton crop.

Table 7
Percentage Area Under Different Crops in Sample Tandas of Maharashtra

(Acres)

District	*Total land of sample tandas (Acres)*	*Cotton*	*Jowar*	*Pulses*	*Sugar-cane*	*Ground-nut*	*Sun-flower*	*Bajra*	*Fruits and veget-ables*	*Others*	*Total*
Beed	6763	17.0	13.5	4.0	–	3.5	5.2	–	–	10	100
Solapur	4791	-	51.7	5.0	–	9.2	18.3	–	–	15.8	100
Osmanabad	2356	2.4	70	5.0	–	3.8	7.5	–	–	11.3	100
Latur	3011	15.0	46.6	9.2	–	0.8	9.2	–	–	19.2	100
Parbhani	8308	34.5	37.7	5.9	–	0.9	1.4	0.9	–	18.7	100
Aurangabad	9227	20.0	30.8	8.0	0.4	0.8	2.3	18.0	0.3	19.3	100
Jalna	6428	32.1	35.0	7.1	–	–	–	9.3	–	16.4	100
Nanded	15600	43.8	35.3	6.0	–	0.3	3.8	0.8	0.3	9.7	100
Buldhana	10962	34.3	30.0	7.1	2.1	–	8.6	–	1.0	16.9	100
Nasik	2146	1.2	20.0	8.8	–	10	–	45.0	–	15.0	100
Akola	11566	47.0	17.5	17.0	–	5.0	–	–	1.0	12.5	100
Jalgaon	6751	19.5	25.0	14.2	–	8.3	–	6.3	3.6	23.1	100
Yeotmal	13977	66.2	17.1	6.8	0.2	–	1.7	–	0.2	7.8	100
Overall	102066	25.62	33.09	8.02	0.21	3.28	8.06	6.18	0.49	15.05	100

Source: Field survey.

The farmers of sample Banjara tandas in Solapur district did not grow cotton crop during the survey year. Further, the farmers of Banjara tandas from Nasik (1.20 per cent) and Osmanabad (2.40 per cent) districts allocated a very little part of their cultivated land to cotton crop during the survey year.

The next important crops grown by the farmers of sample tandas were pulses (8.02 per cent), sunflower (8.06 per cent) and bajra (6.18 per cent). The proportion of area under pulses was relatively higher in the case of tandas in the districts of Latur (9.20 per cent), Aurangabad (8.00 per cent), Nasik (8.80 per cent), Akola (17.00 per cent) and Jalgaon

(14.20 per cent). It means the farmers of these districts in general and that of Jalgaon and Akola districts in particular allocate sufficient lands to pulse crops. The farmers of Banjara tandas of Solapur (18.30 per cent), Latur (9.20 per cent) and Buldhana (8.60 per cent) allocate sufficient land to sunflower crop. The farmers of Banjara tandas of Jalna, Buldhana, Nasik and Akola districts did not grow sunflower crop.

Bajra crop is grown by Banjara mainly in Nasik (45.00 per cent), Aurangabad (18.0 per cent) and Jalna (9.30 per cent) districts on quite a larger area. This crop was not grown by the farmers of Banjara tandas of Beed, Solapur, Osmanabad, Latur, Buldhana, Akola and Yeotmal districts of Maharashtra. The groundnut crop was grown on 3.28 per cent of total cropped area of sample Banjara tandas in Maharashtra. The Banjaras of all the districts in Maharashtra except Jalna, Buldhana and Yeotmal were growing groundnut crop on their lands. The sugarcane crop was found to have grown by the farmers of Banjara tandas in Aurangabad, Buldhana and Yeotmal district sonly. Fruits and vegetables were grown in Banjara tandas belonging to Aurangabad, Nanded, Buldhana, Akola, Jalgaon and Yeotmal districts. Fruits and vegetables were mainly grown in Banjara tandas of Jalgaon, Buldhana and Akola districts of Maharashtra. From the above discussion it is clear that main crop grown by farmers of sample Banjara tandas in Maharashtra were kharif jowar, cotton, sunflower, bajra and pulses. The farmers of very few districts grow sugarcane and fruits and vegetable crops due to the constraint of irrigation facilities available there.

MIGRATION OF BANJARA

The important source of work for the Banjaras in Maharashtra is sugarcane cutting. Most of the Banjara tandas have a very little fertile land and the availability of irrigation water on their lands is also very little. As a result, the farmers residing in Banjara tandas have to grow mostly one crop in a year in kharif season. There is no substantial scope for growing rabi and summer crops on their lands due to these two factors. Most of the Banjara tandas do not depend for a longer period on forest products in the study area. Hence the Banjara families residing in tandas of Maharashtra have to migrate to other places in search of employment. The interviews with the dwellers during the survey period brought out the fact that they migrate to other places for sugarcane cutting with and without bullock carts. They were also found to be migrating to other places for non-sugarcane cutting jobs. The districtwise data about the number of families from sample Banjara tandas migrated for sugarcane cutting with bullock carts during 1997 are presented in Table 8.

It is evident from the table that the percentage of families from Banjara tandas migrated for sugarcane cutting with bullock carts was

12.59 per cent in 1997. It could also be seen from the table that no families from the sample Banjara tandas belonging to Solapur, Parbhani and Akola districts had migrated for sugarcane cutting with bullock carts during the survey year. The Banjaras of these districts do not migrate for sugarcane cutting with bullock carts. The percentage of families from Banjara tandas migrated for sugarcane cutting with bullock carts was found to be very high in Beed (55.60 per cent), Aurangabad (36.54 per cent), Nasik (52.33 per cent) and Jalna (21.22 per cent) districts during 1997. From the above analysis it can be said that the maximum number of families from the Banjara tandas in Beed, Aurangabad, Jalna and Nasik districts do migrate with bullock carts for the work of sugarcane cutting.

Table 8
Seasonal Migration of Banjara Families of Sample Tandas in Maharashtra in 1997

Districts	*Total No. of families*	*For Sugarcane cutting with bullock carts*	*For sugarcane cutting without bullock carts*	*Other than sugarcane cutting*	*Total No. of migrating families*
Beed	723	402 (55.60)	161 (22.27)	90 (12.45)	653 (90.32)
Solapur	815	-	65 (7.98)	53 (6.50)	118 (14.48)
Osmanabad	320	10 (3.13)	57 (17.81)	60 (18.75)	127 (39.69)
Latur	405	04 (0.99)	55 (13.58)	51 (12.59)	110 (27.16)
Parbhani	1182	-	66 (5.58)	150 (12.69)	216 (18.27)
Aurangabad	1248	456 (36.54)	107 (8.57)	44 (3.53)	607 (48.64)
Jalna	542	115 (21.22)	35 (6.46)	66 (12.17)	216 (39.85)
Nanded	1515	15 (0.99)	88 (5.81)	141 (9.31)	244 (16.11)
Buldhana	885	46 (5.20)	127 (14.35)	147 (16.61)	320 (36.16)
Nasik	386	202 (52.33)	30 (7.77)	18 (4.66)	250 (64.76)
Akola	1041	-	61 (5.86)	41 (3.94)	102 (9.80)
Jalgaon	1661	266 (16.01)	95 (5.72)	163 (9.81)	524 (31.54)
Yeotmal	1675	45 (2.69)	181 (10.81)	85 (5.07)	311 (18.57)
Overall	12398	1561 (12.59)	1128 (9.10)	1109 (8.94)	3798 (30.63)

Note: Figures in brackets indicate percentage to total.
Source: Field survey.

Few families of Banjara tandas from Osmanabad, Latur, Nanded, Buldhana, Jalgaon and Yeotmal districts were found to have migrated for sugarcane cutting with bullock carts during 1997. The table shows that the migration of Banjara families for sugarcane cutting with bullock carts was found to be relatively higher in Beed, Aurangabad, Jalna, Nasik and Jalgaon districts. The main reason behind it is that these districts are nearer to Western Maharashtra where sugar factories have started earlier in large numbers compared to other regions. So naturally, Western Maharashtra became the route of migration for Banjaras from the above districts.

LEVEL OF LIVING

The total families of sample Banjara tandas are divided among four different groups according to their economic status or condition, by considering the following criteria.

Well-to-do Families: The well-to-do- families are those families who have sufficient and assured source of income, mainly from land, and have strong economic base and capacity to face adverse economic year, and have capacity to provide employment and credit to others, and are dominant socially and economically in the tanda.

Medium Families: The medium families are those who have assured source of income mainly from land, they have dry and low quality land in a greater proportion, and cannot stand on their own in adverse economic year, and under such condition and for their economic development they have to depend on mainly well-to-do-families.

Poor Families: The poor families are those families who have few pieces of low quality land and get the employment in agriculture and allied activities, including migration, throughout the year.

Very Poor Families: The very poor families are those families who are mainly workers in agriculture and allied activities, and have no guarantee of employment throughout the year.

The districtwise distribution of families in sample tandas according to their economic conditions is presented in Table 9. This table indicates that only 6.07 per cent of the total families in sample Banjara tandas in Maharashtra were economically very sound. The percentage of well-to-do-families was even higher than this in the case of Banjara tandas in Akola, Buldhana, Nanded, Parbhani, Osmanabad (around 8 per cent), Jalna (10.33 per cent) and Aurangabad districts. The percentage of well-to-do-families was very low in the tandas of Nasik, Latur and Solapur districts.

34.47 per cent of the total families in sample Banjara tandas were in medium economic strata. The percentage of such families was higher than this figure in the districts of Akola (54.47 per cent), Buldhana (37.06 per cent), Aurangabad (41.67 per cent), Parbhani (34.94 per cent), Osmanabad (38.13 per cent) and Solapur (36.44 per cent). 36.33 per cent of the total families in sample Banjara tandas were poor families. This percentage was even higher in the tandas of Nasik (50.78 per cent), Nanded (43.43 per cent), Latur (44.94 per cent), Buldhana (40.11 per cent), Osmanabad (40.00 per cent) and Yeotmal (37.55 per cent) districts. The incidence of poverty was relatively higher on the families of Banjara tandas in these districts. At the all-district level the percentage of very poor families was 28.13. It was relatively higher in the case of Banjara tandas in Yeotmal, Jalgaon, Nasik, Jalna, Solapur and Beed districts.

Table 9
Economic Status of Banjara Families in Sample Tandas of Maharashtra

Districts	*Total No. of families*	*No. of well-to-do-families*	*No. of medium families*	*No. of poor families*	*No. of very poor families*
Beed	723	38 (5.26)	217 (30.01)	233 (32.23)	235 (32.50)
Solapur	815	16 (1.97)	297 (36.44)	245 (30.06)	257 (31.53)
Osmanabad	320	25 (7.81)	122 (38.13)	128 (40.00)	45 (14.06)
Latur	405	2 (0.49)	140 (34.57)	182 (44.94)	81 (20.00)
Parbhani	1182	96 (8.12)	413 (34.94)	429 (36.29)	244 (20.65)
Aurangabad	1248	94 (7.53)	520 (41.67)	365 (29.25)	269 (21.55)
Jalna	542	56 (10.33)	153 (28.23)	172 (31.73)	161 (29.71)
Nanded	1515	134 (8.84)	464 (30.63)	658 (43.43)	259 (17.1)
Buldhana	885	71 (8.03)	328 (37.06)	355 (40.11)	131 (14.80)
Nasik	386	1 (0.26)	21 (5.44)	196 (50.78)	168 (43.52)
Akola	1041	84 (8.07)	567 (54.47)	317 (30.45)	73 (7.01)
Jalgaon	1661	40 (2.41)	552 (33.23)	595 (35.82)	474 (28.54)
Yeotmal	1675	96 (5.73)	480 (28.66)	629 (37.55)	470 (28.06)
Overall	12398	753 (6.07)	4274 (34.47)	4504 (36.33)	2867 (23.13)

Note: Figures in brackets indicate percentage to total.
Source: Field survey.

Considering the poor and very poor families together, it seems that there were 59.46 per cent families in sample tandas which were either economically poor or very poor. The districts where the percentage of poor and very poor families was higher than the all district level figure (59.46 per cent) are Nasik (94.30 per cent), Yeotmal (65.61 per cent), Nanded (60.53 per cent), Jalna (61.44 per cent), Latur (64.94 per cent), Solapur (61.59 per cent) and Beed (64.73 per cent). It means that Banjaras of sample tandas in these districts were living in a low economic strata.

CONCLUSION

Banjara people of Maharashtra is that part of Indian society which were engaged in trade and carrying occupation on pack-bullocks from generation to generation, before modern means of transport like railway came into existence. North India is their original home. They migrated to South India with Mughal army for supplying foodgrain to their army camp. After migration they mostly centralized in Deccan (Central India); because Deccan was most suited place for their traditional occupation. Supplying foodgrain to different political powers at the time of war and in time of war interval engag in internal trade were two main aspects of their traditional occupation. For this purpose they were always moving from one part to another. That is why they are known as nomadic tribe in history. They are also known as Criminal tribe. Because like many other tribes they were under criminal Tribes Act 1871 for about eighty years. Today, they are settled down in hilly tracts of Maharashtra, specially Marathwada, Vidarbha and Khandesh regions, and depend directly and indirectly on agriculture. But before settling down on agriculture, they passed through many modes of economic life.

In nomadic time concept of wealth and permanent income sources was not developed among them. But when they lost their traditional occupation, this concept developed among them and they turned to agriculture for this purpose. But the study has clearly indicated that 50 per cent Banjara families of sample tandas were either landless (28.81 per cent) or having land of up to 5 acres. Because of dry and low fertile land Banjaras of Maharashtra do not get employment throughout the year. Seasonal migration, especially sugarcane cutting, is result of it. Dry and low fertile land directly affect on their crop pattern. Finally study indicates that about 60 per cent families of sample Banjara tandas are living under poor condition. Only 6 per cent families found well-to-do.

Uneconomic landholding, absence of full employment, lack of sufficient return to their hardworking and absence of government's proper attention to this denotified community settled in hilly tracts of Maharashtra are the main reasons of their poverty and unemployment and ultimately low level of standard of living.

NOTES & REFERENCES

1. Naik, Ranjit: Report of All India Banjara Study Team, Publisher Study Team, AIBSS, 1968, p. 8.
2. Tavernier, Jean Baptiste: Baron of Aubonne, "Travels in India", translated from the original French edition of 1676, by V.B. Ball, Vol. I, London, Macmillan and Co., and New York, 1889, p. 41.
3. Briggs Captain John: "Account of the Origin, History and Manners of the Race of Men Called Banjaras", published in Transactions of Literary Society of Bombay, Vol. I, London, 1899, pp. 176-77.
4. Bombay: Gazetteer for Hyderabad Assigned Districts, commonly Called Berar, edited by A.C. Lyall, Bombay; printed at the Education Society Press, Byculla, 1870, pp. 196-97.
5. Briggs, Captain John (1819): Op. cit. pp. 185-87.
6. Horatio Bickerstaffe Rowney: "The Wild Tribes of India", B.R. Publishing Corporation, Delhi. Reprinted 1974, First Published 1882, pp. 8-14.
7. Bombay: Gazetteer of the Bombay Presidency, Vol. XVI, Nasik, Printed at the Government Central Press, Bombay, 1883, p. 137.
8. Bombay: Gazette of the Hyderabad Assigned Districts, Commonly called Berar, 1870, Bombay, Printed at the Education Society's Press, Byculla, 1870, p. 196.
9. Briggs, Captain John (1819): Op. cit. pp. 185.

10. Russel R.V. and Hira Lal: "The Tribes and Castes of the central Provinces of India", Vol. II, Rajdhani Book Centre, Delhi, pp. 186-187.
11. Dilip D'Souza, De-notified Tribes: Still 'criminal'? Economic and Political Weekly, December 18-24, 1999, Vol. XXXIV, No. 51, p. 3576.
12. Kaka Kalekar: Report of the Backward Classes Commission, Government of India, Vol. I, Printed in India by the Manager, Government of India Press, Shimla, 1955, p. 36.
13. Munshi K.M.: Report of the Criminal Tribes Act Enquiry Committee, 1939, Bombay, Printed at the Government Central Press, pp. 31-43.
14. Enthoven R.E.: "The Tribes and Castes of Bombay", Vol. I, Cosmo Publication 1975, pp. 271-285.
15. Singh O.P.: "Surat and its trade" in the Second Half of the 17th Century, University of Delhi, 1975, pp. 27-31.
16. Tod James: "Annals and Antiquities of Rajasthan", Vol. II, Published by Brojendar Lall Dass, Calcutta, 1884, pp. 674-677.
17. Enthoven R.E. :"The Tribes and castes of Bombay", Vol. II, Asian Educational Services, New Delhi, Madras, 1990, pp. 332-333.
18. Nanjundayya, H.V. and Ananthakrishna: "Mysore Tribes and Castes", Vol. II, Published under the Auspices of the Mysore University, Mysore, 1928, pp. 174-177.

K.K. Patil

GROWTH PERFORMANCE OF SUGARCANE CROP IN MAHARASHTRA STATE

3

Sugarcane is the most ancient crop of India. Sugarcane information can be found in the ancient epics and other Vedic literature. Historically, eastern India was known to be the centre of origin from where the sugarcane seems to have been carried to China. Subsequently, it was spread to the Philippines, Jawa and other places. Now, it is mostly grown in Cuba, USA, Australia, Philippines, Indonesia, South Africa and Russia.[1]

Sugarcane crop is grown in several rainfall, temperature and soil regions. However, sugarcane growing areas could be grouped into sub-tropical and tropical belt.[2] The States of Maharashtra, Gujarat, Andhra Pradesh, Tamil Nadu and Kerala comprise the 'tropical region' while the States of Punjab, Uttar Pradesh, Bihar and West Bengal comprise the 'sub-tropical' region.

Sugarcane plays a very important role in the agro-industrial economy of India in general and rural economy of important sugar growing States in particular.[3] It provides employment and income to a large number of people. It meets the internal demand for sugar, makes available allied products and earns much needed foreign exchange.

India is the largest producer of sugarcane in the world. There has been considerable importance to this commercial crop during the last two decades.[4] In the year 1995-96 at all India level, the area under sugarcane is 39.21akh hectares and production is 2816 lakh tons with an average yield of 71.8 tons.[5]

Maharashtra is one of the leading sugarcane growing States in India. It is a perennial irrigated crop grown in all the zones of the State. Maharashtra State occupied a well recognized place of pride on the sugarcane map of the country. The area under sugarcane in the Maharashtra State is 5.80 lakh hectares and have total production of 467.05 lakh tons with an average yield 80.526 tons per hectare in the year 1995-96.[6]

The area under sugarcane in Maharashtra is concentrated in low rainfall regions in the eastern part of the sahyadri mountains. Pune division i.e. Ahmednagar, Pune and Solapur districts; Kolhapur division i.e. Satara, Sangli and Kolhapur districts; and Nasik district are the major sugarcane growing districts of the State with 71.96 per cent of the State area. Some districts of Marathwada i.e. Aurangabad and Osmanabad have moderate area under sugarcane i.e. 9.60 per cent of the state area.[7]

Sugarcane cultivation is now, being extended to the other districts of the Marathwada and Vidarbha regions due to the establishment of new sugar factories in these areas and due to the availability of the infrastructure. The present article aims at covering growth performance of sugarcane crop in Maharashtra State. It also attempts to analyse regionwise growth performance of sugarcane crops. For this, four regions viz. Konkan, Western Maharashtra; Marathwada and Vidarbha have been considered.

Triennium average percentage of area and production of sugarcane, for each region, in the Maharashtra State has been calculated. Similarly, regionwise annual compound growth rates of area, production and yield of sugarcane in the Maharashtra State are computed separately for each period. The study period was sub-divided into four periods. Period-I comprises of nine years from 1967-68 to 1995-96; Period-II comprises thirteen years from 1976-77 to 1988-89; Period-III comprises of seven years from 1989-90 to 1995-96; and Period-IV comprises of twentynine years from 1967-68 to 1995-96. For calculating compound annual growth rates the exponential function of the form $y = ab^t$ has been used which becomes a semi-log function of the form $\log y = a + bt$ for the operational purpose. Here, y stands for area, production and yield. 'a' is the intercept, 'b' is the co-efficient and 't' is the time factor. (b-1) x 100 gives per cent growth rate. The standard error of growth rates has been worked out by using the following formula.[8]

$$\mathrm{SE} = \frac{100 \times B}{\log_{10} \theta} \sqrt{\frac{\sum_{1}^{n} \log y^2 - \left(\frac{\sum \log y}{n}\right)^2 - (\log b)^2 \sum_{1}^{n} (t - \bar{t})^2}{(n-2)\sum_{1}^{n} (t - \bar{t})^2}}$$

where, SE (r) - standard error of compound growth rate

B $= r + 1$

$\log_{10} a$ = 0.43429

y = area/production / yield

t = time factor $t = 1 \ldots\ldots\ldots n$

n = No. of observations.

And finally, significance of the growth rates was tested by using students 't' test t i.e.= *rl* SE (r)

t = students 't' distribution with (n - 2) degrees of freedom.

From the table 1 it is clear that the triennium average percentage, for the period 1967-68 to 1969-70, of area under sugarcane for Western Maharashtra is the highest (77.24 per cent) and it is the lowest (00.30 per cent) for Konkan region. The percentage of area under sugarcane for the Marathwada region is 20.20 per cent and for the Vidarbha region it is 2.58 per cent. It can be concluded that the area under sugarcane for the period 1967-68 to 1969-70 is the highest in Western Maharashtra followed by Marathwada.

Table 1
The Trienium Average Percentage of Area and Production of Sugarcane in the Maharashtra State

Period	*Konkan*		*Western Maharashtra*		*Vidarbha*		*Marathwada*		*Maharashtra State*	
	A	*P*	*A*	*P*	*A*	*P*	*A*	*P*	*A*	*P*
1967-68 to	005	291	1300	106040	38	2288	340	18669	1683	127288
1969-70	(00.30)	(0.23)	(77.24)	(83.31)	(2.58)	(1.80)	(20.20)	(14.67)	(100.00)	(100.00)
1976-77 to	005	396	2459	195865	52	3122	584	33889	3100	233271
1978-79	(0.16)	(0.17)	(79.32)	(83.96)	(1.68)	(1.34)	(18.84)	(14.53)	(100.00)	(100.00)
1989-90 to	0.7	63.33	3015	270351	158	11600	1105	79134	4279	361148
1991-92	(0.02)	(0.02)	(70.46)	(74.86)	(3.69)	(3.21)	(25.82)	(21.91)	(100.00)	(100.00)
1993-94 to	00	00	3473	301449	142	10247	1198	84496	4813	396192
1995-96	(00.00)	(00.00)	(72.16)	(76.09)	(2.95)	(2.59)	(24.89)	(21.33)	(100.00)	(100.00)

Note: Bracketed figures indicate the percentages to respective totals.
Source: 1. Season and crop repot, Government of Mahrashtra, 1967-68 to 1974-75.
2. Epitome of agriculture in Maharashtra, 1975-76 to 1994-95.
3. Commissionerate of Agriculture, Pune, 1994-95 to 1995-96.

The triennium average percentage, for the period 1976-77 to 1978-79, of area under sugarcane for Western Maharashtra is the highest (79.32 per cent) and the lowest (0.16 per cent) in the Konkan region. The percentage area under sugarcane for, the Marathwada region is 18.84 per cent and for the Vidarbha region it is 1.68 per cent during the above mentioned period.

For the period 1976-77 to 1978-79, the area under sugarcane was the highest in Western Maharashtra followed by Marathwada.

The triennium average percentage, for the period 1989-90 to 1991-92, of area under sugarcane for Western Maharashtra is the highest (70.46 per cent) and it is the lowest (0.02 per cent) for the Konkan region of the Maharashtra State. The triennium average percentage of area under sugarcane for the Marathwada region is 25.82 per cent and for the Vidarbha region, it is 3.69 per cent. From this it is obvious that the area under sugarcane is the highest in Western Maharashtra and the lowest in - the Konkan region for the period 1989-90 to 1991-92.

The triennium average percentage, for the period 1993-94 to 1995-96, of area under sugarcane for Western Maharashtra is the highest (72.16 per cent) and it is the lowest (00.00 per cent) in the Konkan region of Maharashtra State. The triennium per cent of area under sugarcane for Marathwada is 24.89 per cent and for Vidarbha it is 2.95 per cent. This indicates that the area under sugarcane is the highest in Western Maharashtra followed by Marathwada and Vidarbha. The Konkan region has zero area under sugarcane for this period.

From the above discussion, it is clear that Western Maharashtra is a leading region, in the State of Maharashtra with respect to area under sugarcane. Marathwada ranks second in the state followed by Vidarbha Konkan region has negligible area under sugarcane. The triennium average percentage of production, for the period 1967-68 to 1969-70, of sugarcane for Western Maharashtra is the highest 83.31 per cent and the lowest (00.23 per cent) for Konkan region. The percentage production of sugarcane for Marathwada region is 14.67 per cent and for Vidarbha it is 1.80 per cent. This shows that the production of sugarcane is the highest in Western Maharashtra followed by Marathwada. It is negligible in Vidarbha and the Konkan regions.

The triennium average percentage production for the period 1989-90 to 1991-92, of sugarcane for Western Maharashtra is the highest (i.e. 74.86 per cent). In Vidarbha the annual compound growth rate of sugarcane production was the highest (i.e. 8.50 per cent) and significant at 1 per cent level. This is due to positive and significant growth rate of area under sugarcane. The annual compound growth rate of production of sugarcane was 2.41 per cent (significant at 10 per cent level) for Western Maharashtra; 1.38 per cent for the Marathwada and -8.44 per cent for the Konkan region. The negative and significant growth rate of sugarcane production for the Konkan region was due to the negative growth in area under sugarcane.

The annual compound growth rate of production of sugarcane, for the period III (1989-90 to 1995-96), in the Maharashtra State is 3.63 per cent. It is highest (3.87 per cent) in the Marathwada region followed by Western Maharashtra (3.85 per cent). The lowest production growth rate (1.05 per cent) was noticed during the above mentioned period in the Vidarbha region.

The annual compound growth rate of production of sugarcane, for the 29 years period (i.e. 1967-68 to 1995-96), in the Maharashtra State is 4.27 per cent and it is significant at 1 per cent level indicating substantial increase in the production of sugarcane. This is due to highly significant growth in area (i.e. 00.02 per cent) for the Konkan region. The percentage production of sugarcane for the Marathwada region is 21.91 per cent and for Vidarbha it is 3.21 per cent. It is obvious that the production of sugarcane crop, for this period, is the highest in Western Maharashtra followed by Marathwada and the Konkan regions.

The triennium average percentage production, for the period 1993-94 to 1995-96, of sugar cane for Western Maharashtra is the highest (i.e. 76.09 per cent) and the lowest (i.e. 0.00 per cent) for the Konkan region. The percentage production of sugarcane for the Marathwada region is 21.33 per cent and for Vidarbha is 2.59 per cent. It denotes that the production of sugarcane crop is the highest in Western Maharashtra followed by Marathwada and it is the lowest in Vidarbha. The Konkan region is not having any production of sugarcane during this period.

From the above discussion, it is clear that Western Maharashtra is the leading region with respect to production of sugarcane. The Marathwada region ranks second in the Maharashtra State with respect to sugarcane production. The Vidarbha region ranks third and the Konkan region has negligible production of sugarcane.

REGIONWISE GROWTH PERFORMANCE OR SUGARCANE CROP

Region wise annual compound growth rates of area, production and yield of sugarcane crop in the Maharashtra State are presented in table 2.

From the table 2 it is clear that the annual compound growth rate, for the period 1967-68 to 1975-76, of area under sugarcane in the Maharashtra State 1 is 2.20 per cent which is statistically significant at 10 per cent level. During this period the annual compound growth rate of area under sugarcane was the highest (3.77 per cent) for the Vidarbha region. It was statistically significant at 5 per cent level. Western Maharashtra has 2.91 per cent area growth rate which is statistically significant at 5 per cent level. The annual compound growth rate of area under sugarcane was negative (i.e. -1.18 per cent) for the Marathwada region. The annual compound growth rate of area for the Konkan region in this period is not significant, indicating less importance of the growth of area under sugarcane.

For the period II (1976-77 to 1988-89) the annual compound growth rate of area under sugarcane in the Vidarbha region was the highest (i.e. 6.92 per cent) and it was significant at 1 per cent level. The annual compound growth rates of area under sugarcane in other regions indicate negative growth of area under sugarcane during this period. The growth rates were -0.19 per cent, -0.43 per cent and -8.20 per cent for Marathwada, Western Maharashtra and the Konkan regions respectively. The growth rate of Konkan is negative and signficant at 1 per cent level indicating drastic reduction in the area under sugarcane during the period 1976-77 to 1988-89.

Table 2
Regionwise Annual Compound Growth Rates of Area, Yield and Production of Sugarcane crop in Maharashtra State

(Per cent per annum)

Region		*Period-I 1967-68 to 1975-76 (9 years)*	*Period-II 1975-76 to 1988-89 (13 years)*	*Period-III 1989-90 to 1995-96 (7 years)*	*Period-IV 1967-68 to 1995-96 (29 years)*
1	2	3	4	5	6
Konkan	A	2.56	(-) 8.20***	-	-
		(3.60)	(1.56)		
	P	3.90	(-) 8.44***	-	-
		(3.44)	(0.84)		
	Y	1.67	3.20	-	-
		(1.34)	(4.69)		
Western Maharashtra	A	2.91**	(-) 0.43	4.87*	3.68
		(1.41)	(1.05)	(2.78)	(0.42)
	P	6.84***	2.41*	3.85	3.85***
		(1.58)	(1.35)	(3.35)	(0.37)
	Y	3.82**	1.07*	(-) 0.97	0.04
		(1.90)	(0.66)	(0.92)	(0.26)
Marathwada	A	(-) 1.18	(-) 0.19	4.67	5.71***
		(3.03)	(1.25)	(4.45)	(0.64)
	P	1.00	1.38	3.87	7.05***
		(4.84)	(1.66)	(4.08)	(0.74)
	Y	(-) 1.61	1.58*	(-) 1.29	2.21
		(11.44)	(1.05)	(1.60)	(1.11)
Vidarbha	A	3.77**	6.92	1.34	6.21***
		(1.82)	(1.66)	(5.16)	(0.53)
	P	7.09**	8.50***	1.05	7.24***
		(3.70)	(2.35)	(5.63)	(0.69)
	Y	3.20	1.47	(-) 0.29	0.97***
		(2.24)	(1.43)	(0.90)	(0.35)
Maharashtra State	A	2.20*	(-) 0.23	4.66	4.14***
		(1.44)	(0.96)	(3.19)	(0.41)
	P	6.10***	0.88	3.63	4.27***
		(1.65)	(0.85)	(3.63)	(0.37)
	Y	3.81**	1.11*	(-) 0.55	0.10
		(1.66)	(0.58)	(0.99)	(0.23)

Significant at 10 per cent level*
Significant at 5 per cent level**
Significant at 1 per cent level***

Note: 1. Sugarcane crop area is not reported for Konkan region since 1988-89 onwards.
2. Figures in the brackets indicate standard error of growth rate.

Sources: 1. Season and Crop Report, Government of Maharashtra, 1967-68 to 1974-75.
2. Epitome of Agriculture in Maharashtra, 1975-76 to 1994-95.
3. Commissionerate of Agriculture, Pune, 1994-95 to 1995-96

For the period IIIrd (1989-90 to 1995-96) the annual compound growth rate of area under sugarcane in the Maharashtra State is 4.66 per cent. During this period the annual compound growth rate of area under sugarcane was the highest (i.e. 4.87 per cent) and significant at 10 per cent level for the Western Maharashtra followed by Marathwada (4.67 per cent) and the lowest (i.e. 1.34 per cent) for Vidarbha region.

For the Konkan region, the area under sugarcane has not reported. For the period IV, (1967-68 to 1996-96) the annual compound growth rate of area under sugarcane in the Maharashtra State is 4.14 per cent which is statistically significant at 1 per cent level indicating substantial increase in the area under sugarcane during this period. During this period, the annual compound growth rate of area under sugarcane was the highest (i.e. 6.21 per cent) and significant at 1 per cent level for the Vidarbha region followed by Marathwada (5.71 per cent) and Western Maharashtra (3.6 per cent). The annual compound growth rates for both the regions are statistically significant at 1 per cent level indicating substantial increase in the area under sugarcane during the period 1967-68 to 1995-96.

The annual compound growth rate of sugarcane production for the period 1967-68 to 1975-76 in Maharashtra State is 6.10 per cent which is statistically significant at 1 per cent level. During this period the annual compound growth rate of production of sugarcane was the highest (i.e. 7.09 per cent) and significant at 5 per cent level in the Vidarbha region. This is due to substantial increase in the area under sugarcane. The growth rate of production for the Marathwada region is the lowest (i.e. 1.00 per cent). This is due to negative growth rate of area under sugarcane. The growth rate of sugarcane production for Western Maharashtra is 6.84 per cent and significant at 1 per cent level. This is due to positive and significant growth rate of area under sugarcane. For Konkan the growth rate of production is 3.90 per cent per annum.

The annual compound growth rates of production of sugarcane for period II were 7.24 per cent for Vidarbha, 7.05 per cent for Marathwada and 3.85 per cent for Western Maharashtra. All these growth rates were statistically significant at 1 per cent level indicating substantial increase in the sugarcane production during this period. The growth rates of production for Vidarbha and Marathwada were over and above the growth rate for the State of Maharashtra where as the growth rate of production for Western Maharashtra is below the state level growth rate in the same period.

The yield growth rate of sugarcane, for the period -I (1967-68 to 1975-76), for Maharashtra State is 3.81 per cent which is significant at 5 per cent level. The growth rate of sugarcane yield was the highest (i.e. 3.82 per cent) and significant at 5 per cent level in Western Maharashtra. Vidarbha has yield growth of 3.20 per cent which is significant at 10 per cent level. The growth rate of sugarcane yield is negative (i.e. -1.61 per cent) in the Marathwada. The growth rate of sugarcane yield was 1.67 per cent per annum for the Konkan region.

The yield growth rate of sugarcane for the Maharashtra State in peroid-II (1976-77 to 1988-89) is 1.11 per cent which is significant at 10 per cent level. The growth rate of sugarcane yield was the highest (i.e. 3.20 per cent) in Konkan followed by the yield growth in Western Maharashtra (1.07 per cent). The growth rate in Western Maharashtra is significant at 10 per cent level. The annual growth rate of sugarcane yield was 1.58 per cent and significant at 10 per cent level, in Marathwada region. The yield growth rate for Vidarbha was 1. 47 per cent in the same period.

The annual compound growth rate of yield of sugarcane for the period-III (1989-90 to 1995-96) in Maharashtra State is negative (-0.55 per cent) indicating reduction in the yield level. The annual compound growth rates of yield of sugarcane during this period were negative in all the regions of Maharashtra.

CONCLUSION

Sugarcane is the most ancient crop of India. It plays a very important role in the agro-industrial economy of India in general and rural economy of important sugar growing states in particular. India is the largest producer of sugarcane in world economy. During the last two decades there has been considerable importance to this crop.

Maharashtra is one of the leading sugarcane growing states in India. In the year 1995-96 the area under sugarcane in the Maharashtra State is 5.80 lakh hectares and the production of 467.05 lakh tonnes with an average yield of 0.526 tonnes per hectare. The area under sugarcane in Maharashtra is concentrated in low rainfall regions in the eastern part of the Sahyadri mountains. Pune Division i.e. Ahmednagar, Pune and Solapur districts; and Kolhapur division i.e. Satara, Sangli and Kolhapur districts; and Nasik district are the major sugarcane growing districts of the state with 71.96 per cent of the sugarcane area in the state. Western Maharashtra is a leading region in the state of Maharashtra with respect to area and production of sugarcane crop. Marathwada ranks the second in the Maharashtra state followed by Vidarbha region with respect to area and production of sugarcane crop.

The annual compund growth rate of area under sugarcane crop is the highest in Vidarbha region for the periods I, II and IV. The growth rate of area under sugarcane for the periods-III is the highest in Western Maharashtra. For the period-I the area growth rate is found to be the lowest in the Marathwada region. The growth rate of area under sugarcane is the lowest in Konkan for the periods II, III and IV.

The growth rate of production of sugarcane is the highest in Vidarbha region for the periods I, II and IV. It is the highest in Marathwada for the period-III. The growth rate of production of sugarcane is the lowest in Marathwada for the period-I and in Konkan for the periods II, III and IV.

The growth rate of yield of sugarcane is found to be the highest in Western Maharashtra for the period-I. It is the highest in Konkan for the period-II. The growth rate of sugarcane yield is negative allover the Maharashtra State for the period-III. The growth rate of yield of sugarcane is the highest in Marathwada for the period-IV. The yield growth rates are the lowest in the Marathwada, Western Maharashtra and Konkan for the periods I, II, III and IV respectively. The growth performance of sugarcane, during the overall period (1967-68 to 1995-96), reveals that the sugarcane production has substantially increased in the Vidarbha and Marathwada regions. This is mainly due to increase in the area and the better yield. In Western Maharashtra the production of sugarcane is observed to be comparatively low due to less area and low yield of sugarcane during the same period.

NOTES & REFERENCES

1. Kalyankar, S.P.; Kasle, P.V.; and Nagargoje, S.R.: Con-tribution of Yield and Area to Increase Sugarcane Produc-tion in Maharashtra and India, *Maharashtra Journal of Agricultural Economics,* Vol. 7, No.1, March, 1995.
2. Ibid.
3. Inamke, N.M.; Pokharkar, V.G.; Kamble, S.S. and Nawale, S.K.: Productivity of Sugarcane in Maharashtra: Zonewise Analysis, *Maharashtra Journal of Agricultural Economics,* Vol. 7, No.1, March, 1995.
4. Koranne, V.M.; Nagapure, S.C.; and Ulemale, D.H.: Growth Performance of Sugarcane in Maharashtra State, *Maharashtra Journal of Agricultural Economics ,* Vol. 7, No. 1, March, 1995.
5. Singhal, Vikas: *Hand Book of Indian Agriculture,* Vikas Publishing House, Private Limited, New Delhi, 1995.
6. Chief Statistician: Commissionerate of Agriculture, Pune, *The Districtwise Area. Production and Productivity of Principal Crops in Maharashtra State,* 1994-95 to 1995-96.
7. Koranne,V.M.; Nagapure, S.C.; and Ulemale, D.H.: Op. Cit.
8. Kanhere, R.B.: *Growth and Instability in Qilseeds Crop Economy of Maharashtra,* Unpublished Ph.D. Thesis, Submitted to Dr. Babasaheb Ambedkar Marathwada University, Aurangabad, June, 1997.

M.S. SABLE

ACREAGE RESPONSE OF COMMERCIAL CROPS IN A BACKWARD REASON OF MAHARASHTRA

4

INTRODUCTION

Whether and to what extent Indian farmers' are influenced in their production decisions by economic stimuli is a question of central importance for agricultural planning in India.[1] Basically, there are two opposite views regarding supply response of farmers. One view is that the farmers in under developed countries are not responsive to changes in relative prices. Madan, Neale and olson[2] have found evidence favouring negligible or negative supply response for agricultural output in under developed countries. Other view is that the farmers are no less responsive than those in the developed countries. Raj Krishna[3], in his conclusion on price response, says, "it seems whatever data have been analysed in depth, with well specified models, relative price movements are found to have systematic and significant marginal effects on crop patterns."

Although farmers response to price changes has been the subject of several investigations; the nature and extent to which farmers respond to such changes remains widely debated and controversial issue. If the time series data shows that farmers are responsive to the relative prices, in allocating their resources among different crops they grow, then the desired allocation of inputs could be obtained by using price as an incentive to the farmers; for instance, if a shift toward commercial crop(s) is desired then this could be achieved by raising the relative price(s) of crops fixed by the Agricultural Price Commission (CACP).

Thus, there is no denying the fact that price changes do influence farmers decision making regarding allocation of resources. However, price alone does not determine farmers' acreage allocation decisions as was assumed under the classical static situation qualified by *ceteris paribus* assumptions. But non-price factors are also important in determining farmers' acreage allocation decisions. Therefore, it is hypothesised here that acreage allocation decisions are influenced by price and non-price factors. Since the magnitude of farmers response to different factors varies not only from crop to crop but even for the same crop from district to district due to agro-climatic factors, adoption of technology, institutional and other socio-economic factors. Therefore, such factors must be taken into account while examining response behaviour.

In a vast country like ours, where regional variations in incentives and subsidies could be considerable, knowledge of farmers' acreage response in various regions would provide a better perspective to formulate an appropriate agricultural price policy for growth and stability.

In this paper an attempt has been made to gauge the impact of price and non-price factors on the farmers' decisions to allocate the land to different commercial crops grown in a backward reason of Maharashtra i.e. Marathwada.

THE OBJECTIVES

The following are the main objectives of the present study:

1. To estimate the acreage response of commercial crops viz; cotton, groundnut and sugarcane, grown in Marathwada.
2. To examine the relative importance of price, yield, rainfall, irrigation, lagged acreage, gross return per hectare and trend variable in explaining the variation in acreages under crops studied.
3. To examine the inter-district differences in the supply elasticities among the districts of Marathwada.

THE DATA AND METHODOLOGY

The present study, covers the Marathwada region of Maharashtra State, and is purely based on secondary data at district level. The required time series data, for the period 1967-68 to 1987-88, on area, production, farm harvest prices, gross irrigated area and rainfall during the sowing periods have been extracted from secondary sources published and maintained by the Government of Maharashtra. Season and Crop Reports of Maharashtra State, District wise General Statistical Information of Agricultural Department in Maharashtra, and Socio-Economic Review and District Statistical Abstracts are the main sources of data. For some years the above mentioned data have been obtained from the records maintained by the Directorate of Agriculture, Pune.

THE COMPETING CROPS

While selecting the competing crops, relative importance of the crops, area relationship between or among the crops, types of the crop and season in which it is sown are considered. Cotton and groundnut are considered as the competing crops for sugarcane in Aurangabad, Parbhani and Nanded districts and for Marathwada as a whole, and groundnut as the competing crop for sugarcane in Beed and Osmanabad districts. Kharif Jawar, bajra and groundnut as the competing crops for cotton in Aurgangabad, Beed and Osmanabad districts and for Marathwada as a whole while kharif jawar and groundnut as the competing crops for cotton in Parbhani and Nanded districts. In case of groundnut, kharif jowar, bajra and cotton are considered as the competing crops in Aurgangabd, Beed and Osmanabad districts and for Marathwada as a whole and kharif jowar and cotton as the competing crops for groundnut in Parbhani and Nanded districts.

THE ECONOMETRIC MODEL

In the present study the Nerlovian partial adjustment lag model[4] has been used to measure farmers' response behaviour. More specifically, to study the impact of price variable along with the other non-price variables viz; yield, irrigation, rainfall, lagged acreage, gross return and time variable on farmer's acreage allocation decisions. The simple Nerlovian model assumed a naive expectation of price, implying a simple linear relationship between the planned acreage (A^*_t) and the previous year's price (P_{t-1}). Thus, it is based on the following relations:

$$A^*_t = a + b\,P_{t-1} + U_t \quad \ldots \qquad (I)$$

where, A^*_t = Desired (planned) long run acreage;

P_{t-1} = Previous year's price or one year lagged price;

U_t = Error term and

'a' and 'b' are the constant and co-efficient respectively.

Again, it is related with the actual acreage by assuming that in each period the actual area under crop is adjusted in proportion to the difference between the long run desired level of area and actual area under the crop in question.

$$A_t - A_{t-1} = \beta\,(A^*_t - A_{t-1}) \quad \ldots \qquad (II)$$

$$0 \leq \beta \leq 1$$

where, 'β' is the Nerlovian co-efficient of adjustment.

$0 \leq \beta \leq 1$ meaning that the farmers are able to change the acreage of a crop in any year only to the extent of fraction 'β' i.e. the difference between the acreage they would like to plant and the acreage actually planted in the preceding year, indicator of how fast the farmers are adjusting their expectations. When adjustment is perfect $\beta = 1$. However, In the real world, the value of 'β' lies between 0 and 1.

By substituting the value of A^*_t in equation II we get,

$$A_t = A + B\,P_{t-1} + C\,A_{t-1} + V_t \quad \ldots \qquad (III)$$

where, $A = a\beta$, $B = b\,\beta$, $C = 1 - \beta$ and $V_t = \beta U_t$

The reduced form of equation would remain basically the same even if we include more independent variables than the one included in the equation III.

Using the above model as the basic frame of analysis, by considering sown area as proxy to supply which is dependent variable, and including the lagged relative price, lagged relative yield, gross irrigated area lagged one year, rainfall, lagged acreage, gross return per hectare and time trend as the independent variables the response relationship is estimated with the help of following alternative equations:

$$A^*_t = a + b\,P_{t-1} + c\,Y_{t-1} + d\,I_{t-1} + e\,w_t + f\,GR_{t-1} + hT + U_t \ldots \quad (I)$$

where, A^*_t is the area that farmers would plant in the year 't', if there were no difficulties of adjustments to changes in the price and other factors;

P_{t-1} is the relative price of the crop i.e. the farm harvest price of the crop deflated by the weighted price of the competing crops, lagged one year;

Y_{t-1} is the relative yield of the crop i.e. the yield of the crop deflated by the weighted yield of the competing crops, lagged one year;

I_{t-1} is the gross irrigated area under all crops, lagged one year;

W_t is the rainfall (in mm) during the sowing time period;

GR_{t-1} is the gross return per hectare, lagged one year;

T is the trend variable which explains how the output is changing over a time period;

U_t is the error term which explains the proportion of variation in the dependent variable not accounted for by the independent variables included in the regression function;

a is the constant; and

b,c,d,e,f,g and h are the co-efficients

$$A_t - A_{t-1} = \beta\,(A^*_t - A_{t-1}) \ldots \quad (II)$$

$$0 \leq \beta \leq 1$$

where, A_t is the actual area planted under the crop studied in time 't-1

'β' is the co-efficient of adjustment.

by solving the equation I and II we get:

$$A_t = a0 + b_1 P_{t-1} + b_2 Y_{t-1} + b_3 I_{t-1} + b_4 w_t + b_5 A_{t-1} + b_6 GR_{t-1} + b_7 T + V_t \quad (III)$$

where, $a0 = a\,\beta$, $b_1 = b\,\beta$, $b_2 = c\,\beta$, $b_3 = d\,\beta$, $b_4 = e\,\beta$, $b_5 = 1-\beta$, $b_6 = f\,\beta$, $b_7 = h\,\beta$, and $V_t = \beta\,U_t$

To get the elasticities directly, we have estimated the above equation using the ordinary least squares method with all variables in their log-linear form. The short run elasticity (SRE) and the long run elasticity (LRE) are obtained by using the following relations:

$$\text{Short run price elasticity of acreage} = \frac{\text{Mean value of the Independent variable}}{\text{Mean Value of the dependent variable}} \times \text{Co-efficient of price variable}$$

$$\text{Long run price elasticity of acreage} = \frac{\text{Short run price elasticity}}{\text{Co-efficient of adjustment}}$$

RESULTS AND DISCUSSIONS

The estimated acreage response co-efficients for sugarcane, cotton and groundnut in different districts of Marathwada and for Marathwada region as a whole alongwith the value of co-efficient of multiple determination (R^2), the co-efficient of adjustment (β) and standard errors are presented in the tables 1 to 3

Table 1
Estimated Acreage Response Co-efficients for Sugarcane in Different Districts of Marathwada

	Aurangabad	*Parbhani*	*Beed*	*Nanded*	*Osmana-bad*	*Marath wada*
Constant in Log terms (a)	-6.0232	1.3796	2.8504	12.9108	-42.3750	-5.9748
Relative Price Lagged One Year (P_{t-1})	-0.0694 (0.1889)	0.1593 (0.1670)	-6.2303 (22.4013)	-1.1449 (12.6363)	-13.7149 (57.6418)	0.2104 (0.1764)
Relative Price Lagged Two Year (P_{t-2})	-0.0466 (0.1068)	0.0010 (0.1048)	2.3419 (14.2117)	2.2511 (8.9369)	-9.6805 (30.6151)	0.0229 (0.0923)
Relative Yield Lagged One Year (Y_{t-1})	-0.0506 (0.0906)	-0.0591 (0.0799)	0.3598 (0.6168)	-0.1425 (0.3250)	1.1781 (1.1043)	0.1203 (0.0727)
Gross Irrigated area Lagged One Year (I_{t-1})	0.9218[c] (0.2790)	0.5197[b] (0.1780)	0.192 (0.0214)	0.0180 (0.0618)	0.0976[b] (0.0379)	1.0699[c] (0.2696)
Sowing Period Rainfall (W_t)	0.1035 (0.1653)	-0.1789 (0.1130)	-0.0190 (0.0213)	0.0055 (0.0151)	-0.0189 (0.0427)	0.0530 (0.1316)
Lagged Acreage (A_{t-1})	0.0683 (0.2530)	0.3454 (0.2318)	0.4576 (0.3590)	0.4301 (0.3655)	0.3568 (0.2705)	0.2426 (0.2135)
Gross Return Lagged One Year (GR_{t-1})	0.4220[b] (0.1664)	-0.0519 (0.1030)	0.0014 (0.0012)	0.0001 (0.0014)	0.0052 (0.0030)	0.1481 (0.1382)
Time Trend (T_t)	-0.2078[a] (0.1003)	-0.0928 (0.1042)	0.2690 (1.5739)	0.6990 (1.6153)	-4.0283 (3.0757)	-0.1780[b] (0.0778)
Coefficient of Multiple deter-mination (R^2)	0.8944	0.8933	0.8381	0.6088	0.8090	0.9449

Contd....

	Aurangabad	*Parbhani*	*Beed*	*Nanded*	*Osmanabad*	*Marathwada*
Coefficient of Adjustment (β)	0.9317	0.6546	0.5424	0.5699	0.6432	0.7574
Long run acreage elasticity w.r. to price	-0.0500	0.0015	4.3176	3.9499	-15.0505	0.0302

a-Significant at 10 per cent level

b-Significant at 5 per cent level

c-Significant at 1 per cent level

Note: 1. Figures in parentheses are standard errors.

2. Long run acreage elasticities are with respect to p_{t-2}

THE ESTIMATED ACREAGE RESPONSE CO-EFFICIENT FOR SUGARCANE

Relative Price Lagged One Year

The regression co-efficient of relative price variable (P_{t-1}) is found to be positive in case of Parbhani districts and for Marathwada region as a whole. But it is found to be negative in case of Aurangabad, Beed, Nanded and Osmanabad district. None of the co-efficients seem to be statistically significant.

Relative Price Lagged Two year

The co-efficient of relative price lagged two year (P_{t-2}) is found to be positive in case of Parbhani, Beed and Nanded districts and for Marathwada region as a whole. Raj Krishna[5] also found positive price co-efficients for sugarcane in Punjab. Jawahar Kaul also found positive price co-efficient in case of sugarcane for Punjab and Madhya Pradesh.[6] M.C. Madhavan also found similar results in Madras.[7] The regression co-efficient of this variable is found to be negative in case of Aurangabad and Osmanabad districts. None of the co-efficients, either positive or negative, are found to be statistically significant.

Relative Yield Lagged One Year

The relative yield variable seemed to have a positive bearing on acreage allocation decisions in case of Beed and Osmanabad district and for Marathwada as a whole. And negative influence in case of Aurangabad, Parbhani and Nanded districts. Both the positive and negative co-efficients are found to be statistically insignificant.

Gross Irrigated Area Lagged One Year

The estimated parameter with respect to this variable is positive and highly significant at one per cent level in case of Aurangabad district and for Marathwada as a whole. It is positive and significant at 5 per cent level in case of Parbhani and Osmanabad districts and positive but in significant in case of Beed and Nanded districts. This reveals that irrigation factor is one of the most important factors in determining sugarcane acreage though the degree of responsiveness varied substantially from district to district.

Sowing Period Rainfall

The co-efficient of sowing period rainfall seems to be positive, but not significant, in case of Aurangabad and Nanded districts and for Marathwada as a whole. It is negative and in significant in case of Parbhani, Beed & Osmanabad districts.

Lagged Acreage

The elasticity estimates of one year lagged sugarcane acreage are observed to be consistently positive in all the districts and for Marathwada as a whole. But the influence of this variable on acreage allocation decision is not so strong

Gross Return Lagged One Year

The gross return variable has exerted positive and significant impact on acreage allocation decisions in case of Aurangabd district. The regression co-efficient of this variable in Aurangabd district is positive and statistically significant at 5 per cent level with an elasticity of 0.4220 and standard error 0.1664. The regression co-efficient of this variable is also positive in case of Beed, Nanded and Osmanabad districts and for Marathwada as a whole. The positive and significant parameter estimates explain the fact that the farmers do start with taking advantage of gross return as they become more and more aware of the profitability by taking into account the relative price and yield and/or profitability in question. The co-efficient for Parbhani district is negative and statistically insignificant.

Time Trend

The trend variable exerted positive impact only in case of Beed and Nanded districts. It is negative in all other cases.

Co-efficient of Multiple Determination (R^2)

The value of R^2 for sugarcane in case of Marathwada is 0.9449. Among the districts it is highest (0.8944) in case of Aurangabad district and lowest (0.6088) in case of Osmanabad district. This shows that 60 per cent and 89 per cent of variation in sugarcane acreage is explained by the effect of included independent variables in case of Nanded and Aurangabad districts respectively.

Adjustment Co-efficients, Short and Long Run Elasticities

The short and long run acreage elasticities with respect to price show a considerable variation among the districts. The short run acreage elasticity with respect to price of sugarcane lagged two year for Marathwada is 0.0229 which is positive. The long run acreage elasticity with respect to relative price of sugarcane lagged two year for Marathwada is 0.0302. It indicates that the farmers are less responsive to prices in the short run but they are relatively more responsive in the long run. The short run acreage elasticity with respect to price is positive and highest (02.3419) in Beed district indicating a comparatively greater percentage of area under sugarcane is affected by changes in relative price of sugarcane. It is negative and very high (-9.6805) in Osmanabad district indicating greater percentage of area under sugarcane is negatively affected by changes in the relative price of sugarcane. The long run acreage elasticity with respect to relative price lagged two year is positive and highest (4.3176) in Beed district and negative and very high (-15.0505) in Osmanabad district. The co-efficient of adjustment with respect to sugarcane varies from 0.5424 in Beed district to 0.9317 in Aurangabad district. The highest co-efficient of adjustment in Aurangabad district indicates that the desired adjustments in acreage are more rapidly made.

Table 2
Estimated Acreage Response Co-efficients for Cotton in Different Districts of Marathwada

	Aurangabad	*Parbhani*	*Beed*	*Nanded*	*Osmana-bad*	*Marathwada*
Constant in Log terms (a)	2.1480	3.1287	-0.5200	4.5622	4.8662	1.4809
Relative Price Lagged One Year (P_{t-1})	0.3987[a] (0.2074)	0.0942 (0.1303)	-0.8272[a] (0.3930)	-0.0223 (0.0606)	-0.0069 (0.2388)	0.1158 (0.1042)
Relative Yield Lagged One Year (Y_{t-1})	-0.0384 (0.0747)	0.0040 (0.0457)	0.1710 (0.2505)	-0.0436 (0.0514)	0.0107 (0.1209)	0.1691[a] (0.0940)
Gross Irrigated area Lagged One Year (I_{t-1})	-0.0903 (0.1756)	-0.0023 (0.1488)	0.7724[a] (0.3787)	0.0486 (0.0589)	-0.1124 (0.1712)	0.3418[b] (0.1498)
Sowing Period Rainfall (W_t)	0.1621 (0.1052)	0.1064 (0.1065)	-0.0520 (0.3068)	-0.0427 (0.0367)	0.0099 (0.1403)	-0.0674 (0.0777)
Lagged Acreage (A_{t-1})	0.6419[c] (0.1400)	0.4909[a] (0.2327)	0.1716 (0.1912)	0.3325[a] (0.1780)	0.2708 (0.2041)	0.6246[c] (0.1294)
Gross Return Lagged One Year (GR_{t-1})	0.0120 (1062)	0.0435 (0.0478)	0.3187 (0.2338)	0.0953[b] (0.0387)	0.0924 (0.0963)	-0.0207 (0.0711)
Time Trend (T_t)	-0.0786 (0.0674)	-0.0438 (0.0859)	-0.3033[a] (0.1579)	-0.0527[a] (0.0255)	-0.1995[a] (0.1046)	-0.1270[b] (0.0439)

Contd....

	Aurangabad	*Parbhani*	*Beed*	*Nanded*	*Osmana-bad*	*Marath wada*
Coefficient of Multiple determination (R^2)	0.8122	0.5600	0.6073	0.8103	0.7858	0.8325
Coefficient of Adjustment (β)	0.3581	0.5091	0.8284	0.6675	0.7292	0.3754
Long run acreage elasticity w.r. to price	1.1133	0.1850	-0.9985	-0.0334	-0.0094	0.3084

a- Significant at 10 per cent level.
b- Significant at 5 per cent level.
c- Significant at 1 per cent level.
Note: Figures in parentheses are standard errors.

THE ESTIMATED ACREAGE RESPONSE CO-EFFICIENTS FOR COTTON

Relative Price Lagged One Year

The regression co-efficient of this variable is found to be positive, but not significant, for Marathwada as a whole. It is found to be positive in Aurangabad (significant at 10 per cent level) and Parbhani districts also. This indicates that the relative price of cotton has shown significant influence on acreage allocation decisions. Similar result was also found in several studies.[8] But the co-efficients of price variable have negative sign in case of Beed, Nanded and Osmanabad distircts indicating negative and insignificant influence in determining the acreage the under cotton. Several past studies have indicated negative impact of relative price variable on acreage under cotton.[9] This might be due to decline in the cotton productivity per hectare and on the other hand relative increase in that of the competing crops.

Relative Yield Lagged One Year

The relative yield variable is observed to have a positive bearing on acreage allocation decisions in case of Parbhani, Beed & Osmanabad districts and for Marathwad as a whole. However, the co-efficient of relative yield variable is found to be statistically significant at 10 per cent level only in case of Marathwada as a whole. The non-significance of the impact observed in Aurangabad and Nanded districts could not reveal any definite consequence of the yield factor in area allocation decisions.

Gross Irrigated Area Lagged One Year

The regression co-efficient of irrigation is found to be positive in case of Beed and Nanded districts and for Marathwada as a whole. This indicates positive influence on area allocation decisions. The co-efficients of this variable are found to be negative in case of Aurangabad, Parbhani and Osmanabad districts.

Sowing Period Rainfall

The regression co-efficients of this variable are observed to be positive in Aurangabad, Parbhani and Osmanabad districts while negative in case of Beed and Nanded districts and for Marathwada as a whole.

Lagged Acreage

The co-efficient of lagged acreage variable is found to be positive in case of all the districts and for Marathwada as a whole. It indicates that the lagged acreage variable has a positive impact on acreage decisions with respect to cotton crop.

Gross Return Lagged One Year

The regression co-efficient of this variable is found to be positive in all the districts but it is significant at 5 per cent level in case of Nanded district only. It is negative and insignificant for Marathwada as a whole.

Time Trend

The regression co-efficient of trend variable is turned out to be negative in almost all the cases indicating downward revision of the acreage allocation to cotton in these districts.

Co-efficient of Multiple Determination

The value of R^2 for cotton in Marathwada region is 0.8325. Among the districts it is highest (0.8122) in Aurangabad district as compared to other districts and lowest (0.5600) in Parbhani district.

Adjustment Co-efficients, Short and Long Run Elasticities

The short run and long run acreage elasticities with respect to price have shown a considerable variation among the districts. The short run acreage elasticity with respect to price is negative and quite high (-0.8272) in Beed district indicating a large percentage of cotton area is affected by changes in the price of cotton. The highest (0.3987) and positive short run elasticity is found in Aurangabad district. The magnitude of long run elasticity is observed to be in the range of (-0.9985) in Beed district to 1.1133 in Aurangabad district. In respect of cotton the co-efficient of adjustment varies from 0.3581 in Aurangabad district to 0.8284 in Beed district. The highest co-efficient of adjustment for Beed district indicates that the desired adjustment in cotton acreage are more rapidly made and nor-price factors rather than relative price variable operated more quickly. The lowest rate of adjustment in Aurangabad district indicates that the acreage under cotton is influenced more by technological and institutional rigidities and price inducement influenced a little.

Table 3
Estimated Acreage Response Co-efficients for Groundnut in Different Districts of Marathwada

	Aurangabad	*Parbhani*	*Beed*	*Nanded*	*Osmana-bad*	*Marathwada*
Constant in Log terms (a)	3.9912	3.3731	4.2598	5.9503	3.0258	4.2135
Relative Price Lagged One Year (P_{t-1})	-0.5206[b] (0.2141)	-0.0531 (0.1491)	-0.5008 (0.4481)	0.1481 (0.3818)	-0.1211 (0.1678)	0.0817 (0.3059)
Relative Yield Lagged One Year (Y_{t-1})	-0.3044[b] (0.1131)	-0.0708 (0.2653)	0.0812 (0.2281)	0.1035 (0.1982)	0.0162 (0.0987)	0.1662 (0.1914)
Gross Irrigated area Lagged One Year (I_{t-1})	-0.2117 (0.1750)	-0.0332 (0.6308)	-0.0205 (0.2966)	-0.8900[b] (0.3483)	-0.1649 (0.1630)	-0.0896 (0.2741)
Sowing Period Rainfall (W_t)	-0.0205 (0.0825)	0.3657 (0.3514)	0.4114[c] (0.1342)	0.5055[a] (0.2812)	-0.0170 (0.1063)	0.2241 (0.1637)
Lagged Acreage (A_{t-1})	0.4272 (0.2576)	0.0397 (0.2985)	0.2128 (0.2049)	0.3224 (0.1921)	0.6982[c] (0.2247)	0.4768[a] (0.2261)
Gross Return Lagged One Year (GR_{t-1})	0.2239[a] (0.1111)	-0.1506 (0.533)	-0.1796 (0.2364)	-0.0718 (0.2077)	0.0679 (0.0823)	-0.1196 (0.1599)
Time Trend (T_t)	-0.2130[a] (0.1147)	0.2972 (0.3000)	-0.2269 (0.1660)	0.1831 (0.1622)	-0.0937 (0.1162)	-0.0303 (0.1236)
Coefficient of Multiple determination (R^2)	0.7907	0.3812	0.8660	0.8137	0.9385	0.8840
Coefficient of Adjustment (β)	0.5728	0.9603	0.7872	0.6776	0.3018	0.5232
Long run acreage elasticity w.r. to price	-0.9088	-0.0552	0.6361	0.2185	-0.4012	0.1561

a- Significant at 10 per cent level
b- Significant at 5 per cent level
c- Significant at 1 per cent level
Note: Figures in parentheses are standard errors

THE ESTIMATED ACREAGE RESPONSE CO-EFFICIENTS FOR GROUNDNUT

Relative Price Lagged One Year

The regression co-efficient of relative price variable is found to be positive for Nanded district and for Marathwada as a whole. Studies conducted for groundnut by Reddy, Jhay, Kaul and Madhavan had also found the same result.[10] But these co-efficients are statistically insignificant. The district level picture reveals that there is a negative impact of price variable in case of Aurangabad, Parbhani, Beed and Osmanabad districts.

Relative Yield Lagged One Year

The regression co-efficients with respect to relative yield are found to be positive, but not significant, in case of Beed, Nanded and Osmanabad and for Marathwada as a whole. But are found to be negative in case of Aurangabad and Parbhani districts.

Gross Irrigated area Lagged One Year

The gross irrigated area has exerted a negative impact on acreage allocation decisions in all the cases. The negative sign in case of Nanded district, as is indicated, is significant at 5 per cent level. It explains that with the expansion of irrigation facilities the area under groundnut has declined.

Sowing Period Rainfall

The rainfall received during the sowing period has shown positive impact on acreage allocation decisions with respect to groundnut in case of Beed and Nanded districts and for Marathwada as a whole. In the remaining cases the influence of rainfall is negative but not significant.

Lagged Acreage Variable

The lagged acreage variable has emerged as one of the important determinants of acreage under groundnut in all the districts and for Marathwada as a whole. It has positive and significant influence on acreage under groundnut in case of Osmanabad district and for Marathwada region as a whole. In other cases the impact is positive but not significant.

Gross Return Lagged One Year

The gross return variable is observed to have a positive impact on the acreage allocation decisions in respect of Aurangabad and Osmanabad distritcts but the co-efficient is significant at 10 per cent level in case of Aurangabad district only, indicating more acreage to groundnut crop. The negative co-efficients are observed in case of Parbhani, Beed and Nanded districts and for Marathwada as a whole. However, all the negative co-efficients are not significant.

Time Trend

The estimated co-efficient of trend variable with respect to groundnut is positive in case of Parbhani and Nanded districts but they are statistically insignificant. The co-efficient has negative sign in case of Aurangabad, Beed and Osmanabad districts and for Marathwada as a whole. It indicates downward shift in area under groundnut with the passage of time.

The Co-efficient of Multiple Determination

The value of R^2 with respect to groundnut for Marathwada region is 0.8840. The value of R^2 ranged from 0.3812 for Parbhani district to 0.9385 for Osmanabad district.

Adjustment Co-efficents, Short and Long Run Elasticities

The short run and long run acreage elascticities with respect to price show a considerable variation among the districts. The short run acreage elasticity for Marathwada is 0.0817. But it is negative and very low (-0.0531) in case of Parbhani district which indicates that only a small percentage of area under groundnut is affected by changes in the price of groundnut. The highest (0.1481) short run acreage elasticity with respect to price is found in case of Nanded district indicating a comparatively greater percentage of area under groundnut is affected by changes in the price of groundnut. The co-efficient of adjustment with respect to groundnut crop varies from 0.3018 in Osmanabad district to 0.9603 in Parbhani district. The co-efficient of adjustment for Marathwada is found to be 0.5232. The highest co-efficient of adjustment in Parbhani district indicates that the desired adjustments in acreage are more rapidly made and non-price factors rather than relative price operated more quickly.

CONCLUSION AND POLICY MEASURES

The relative price lagged two year is an important factor rather than one year lagged price in affecting sugarcane acreage in most of the districts of Marathwada. The irrigation is considered as the most important factor in determining the sugarcane acreage rather than price and yield. If substantial expansion in sugarcane acreage and production is expected then emphasis should be given to increase irrigation facilities. Under utilization of the existing irrigation potential is another problem of this region. Therefore, training should be given to the farmers regarding water management which would help to raise utilisation ratio in the region.

The acreage under cotton is positively affected by the relative price, relative yield, irrigation, lagged acreage and gross return variable. The emphasis needs to be given on raising the cotton yield and in increasing irrigated area under cotton in Marathwada. Incentives needs to be given to the farmers to grow cotton on their irrigated land and water rates should be made relatively cheaper for cotton. Yield raising measures should be adopted alongwith suitable price and marketing polices in case of cotton to raise the production and income from this crop.

The lagged acreage, rainfall, relative price and relative yield are found to be the most significant factors affecting acreage under groundnut in this region. The measures which ensure higher gross return per hectare from groundnut needs to be adopted. All the yield raising measures like fertilizers, new methods of cultivation and new technology should be adopted along with appropriate price policy to stimulate groundnut production in this region. There is an urgent need for improving production technology of this crop to yield higher return through higher productivity as increased prices alone in the absence of any viable technology may not yield the desired results.

NOTES & REFERENCES

1. Bhagawati, J. N. and Chakarvarthy, S.: "Contribution to Indian Economic Analysis–A Survey", *The American Economic Review,* Vol. LIX, No. 4, Part-2, September-1969, Supplement.
2. Madan, B.K.: "Presidential Address", *Proceedings of the Eighteenth Conference of the Indian Society of Agricultural Economics,* 1958.
3. Krishna, Raj: "Rapporteur's Report on Economics of the Cropping Pattern", *Indian Journal of Agricultural Economics,*Vol XVIII, No. 1, January-March,1963.
4. Nerlove, Marc: *The Dynamics of Supply : Estimation of Farmers' Response to Price,* John Hopkins Press Baltimore,1958.
5. Krishna, Raj: "Farm Supply Response in India-Pakistan: A Case Study of the Punjab Region." *Economic Journal*, Vol. LXXIII, No. 291, September, 1963.
6. Kaul, J.L.: "A Study of Supply Responses to Price of Punjab Crops", *Indian Journal of Economics,* Vol. 48, July, 1967.
7. Madhavan, M.C.: "Acreage Response of Indian Farmers: A Case Study of Tamilnadu", *Indian Journal of Agricultural Economics,* Vol. XXVII, No. 1, January-March, 1972.
8. Sidhu, J.S. and Sidhu, R.S.: "Growth and Area Response of Commercial Crops in Punjab: Some Policy Issues", *Indian Journal of Agricultural Economics,* Vol. 43, No. 3, July-September, 1988.
9. Cummings, J.T.: "The Supply Responsiveness of Indian Farmers in the post-independence period: Major cereals and cash crops", *Indian Journal of Agricultural Economics,*Vol XXX, No. 1, January-March, 1975.
10. Reddy, J.M.: "Estimation of Farmers Supply Response-A Case Study of Groundnut", *Indian Journal of Agricultural Economics,* Vol. XXV, 1970.
11. Jhal, M.L.: "Farmers' Response to Economic Incentives: An Analysis of Inter-regional Groundnut Supply Response in India", *Indian Journal of Agricultural Economics,*Vol. XXXIV, No. 1, January-March, 1979.
12. Kaul, J.L. and Sidhu, D.S.: "Acreage Response to Prices for Major Crops in Punjab-An Econometric Study," *Indian Journal of Agricultural Economics,*Vol XXVI, No. 4, October-December, 1971.

A.N. Pawar

IRRIGATION DEVELOPMENT IN MAHARASHTRA

5

POSITION OF IRRIGATION DEVELOPMENT IN INDIA

Irrigation has played a vital role in continuous process of agricultural development in India. In successive five year plans irrigation development has been given top priority. During the pre- independence period, the emphasis used to be given for developing water resources only as a protective measure against crop failure and famine. In the post independence era , emphasis was changed to develop irrigation as one of the main productive inputs in agriculture.

At the time of commencement of the first five year plan in 1951, population of India was about 361 million and annual food grain production was 51 million, which was not adequate. Import of food grains was then inevitable to cover up the shortage. Attaining self- sufficiency in food was therefore given paramount importance in the plan period and in order to achieve the objective, various irrigation projects were formulated and implemented through successive five year plans to create additional irrigation potential throughout country. This drive compounded with green revolution in the agricultural sector has enabled India to become marginally surplus from deficit one in food grains. In 1996-97 food grains production in the country has been 198 million tones for a population of about 900 million.[1]

With the massive investment in irrigation schemes, it became possible to undertake 292 major and 944 medium projects during last 47 years (i.e. 1951-1997) in India. Thus, the irrigation facility in India rose from 22.60 million hectares to 79 million hectares during last 47 years. As compared with just about 18 per cent of the cropped area under irrigation at the time of independence, the irrigation level has since increased to 38 per cent in 1995-96 though there has been an increase of nearly 42 per cent in cropped area.

IRRIGATION DEVELOPMENT IN MAHARASHTRA STATE

The linguistic state of Maharashtra was politically evolved on the first of May 1960 as a late consequence of the states, reorganization schemes of the Government of India. Commanding a more or less central situation in the Republic of India, Maharashtra is the third largest state in republic both in population and in area.[2] The total geographical area of the state is 307.58 lakh hectares and the net sown area of 179.11 lakh ha. (58.23 per cent of geographical area). The gross cropped area in the state comes to 218.97 lakh hectares and the percentage of gross irrigated area to gross cropped area is found to be 15.40 per cent by the end of the year 1999-2000.[3]

The main feature of the Maharashtra state is that agriculture continues to be the major source of income of the state. The distinctive feature of the state is that it is situated in a relatively low rainfall tract, 33 per cent of the state area is the drought prone. The rainfall in the state varies from 700 cm in the Konkan area (coastal belt) to 50 cm in the drought prone areas. The rainy days are 45 in a year. The lack of adequate rain underlines the need for irrigation. Culture and history of irrigation also indicates that civilization in the state has developed and flourished on the banks of major rivers in the state like Godavari, Krishna, Bhima, Panchganga, Nira, Tapi, Purna, Wainganga, Painganga etc. Agriculture of Maharashtra state is to a large, unevenly distributed among the regions in Maharashtra. About 33 per cent of population of Maharashtra state is suffering from scarcity.[4]

Irrigation development in Maharashtra gained momentum after the formation of the state. Thus, before the formation of state i.e. up to 1960, the achievement on the irrigation front in Maharashtra on the whole remained considerably poor. The gross irrigated area of the state was amounted to a mere 6 per cent of gross cropped area and irrigation potential in the state amounted to a mere 2.16 per cent of net sown area in 1960. The efforts for the development of irrigation in state stared since the formation of Maharashtra state in 1960.[5] The first and immediate action of the Maharashtra Government was the appointment of state irrigation Commission. This commission made a comprehensive study of irrigation potential in the state and submitted its report to government in 1962. This commission hoped that near about 30 per cent cropped area can be irrigated by river water and after completion of all proposed large, medium and minor irrigation projects.

Water Resources in Maharashtra

In Maharashtra, five major river basins viz. Godavari, Krishna, Tapi, Narmada and Konkan constitute the major water resources. The basin wise distribution of available water is shown in the table 1 and it is seen from this table that, distribution of total water being 31.06 per cent, 20.77 per cent, 5.57 per cent, 0.35 per cent and 42.25 per cent in Godavari, Krishna, Tapi, Narmada and Konkan river basins, respectively.

Government of Maharashtra had recently established second Irrigation Commission under the chairmanship of Dr. M.A. Chitale in December 1995. This commission has submitted its report to the state government in October 1999 and made comprehensive study of the irrigation in Maharashtra. The commission has estimated that 9.7 mha. is to be the ultimate irrigation potential for the state and out of which 4.1 mha. to be from the ground water and 5.6 mha. to be from the surface water. This commission has also expected that, by using the modern irrigation technique and improving water management distribution land, the ultimate irrigation potential in the state would be increased up to 12.6 mha.

Potential and Utilizable Water Resources

Potential and utilizable water resources for the state have been recently assessed by the Second Irrigation. Commission (1995) and is presented in table 1. Table 1 reveals that, average run-off in the river system of the state has been assessed as 1,63,820 million cubic meters (163.82 cubic kilometers) while 1,869 cubic kilometers has been estimated for the country and it is9 per cent of the estimate of the country. Based on 1991 census per capital availability of water in the state workout to 2076 cubic meters and 2200 cubic meters for the country in year 1997. But due to limit of the inter-state water distribution 1,25,936 million cubic meters (77 per cent of total potential) would be utilisable water available in the state.

Utilization of Water for Different Purposes (In the Year 1996 and 2030 A.D.)

Second Irrigation Commission has worked out a plan about the utilization of the water for different purposes in the year 1996 and in the year 2030 A.D. It is seen from the table 2 that the total of 39484 million cubic meter water has been utilized in the year 1996 for the different purposes. Out of this, 79 per cent was utilized for irrigation purpose and remaining 21 per cent for other purposes viz. (i.e. 9 per cent for drinking and 12 per cent for industry). This fact shows that only 28.3 per cent of the total available water was utilized for different purposes in this year (i.e. 22.4 per cent for irrigation and 5.9 per cent for other purposes).

Demand for water in future period would be always increasing due to increase in population. It is also seen from this table that in the year 2030 A.D. availability of water in the state will be 1,39,227 million cubic

Table 1
Basin-wise Water Availability in Maharashtra

Geographical area : Thousand sq. km.
Cultivated area : Lakh hactares
Water : Million cubic meters

Sr. No.	*River basin*	*Total geographical area covered by the basin*	*Percentage of basin area covered to total geographical area (of 3)*	*Cultivable area*	*Percentage of cultivable area to total geographical area (of 5)*	*Average water availability*	*Permission for utilization of water*	*Percentage of permission to level of availability*
1	2	3	4	5	6	7	8	9
1.	Godavari	152.81	46.71	112.56	49.94	50880	34185	67.19
2.	Krishna	69.03	22.45	56.27	24.96	34032	16818	49.42
3.	Tapi	51.3	16.69	37.31	16.55	9118	5415	59.39
4.	Narmada	1.05	0.34	0.06	0.28	580	308	53.1
5.	Konkan	33.23	10.91	18.64	8.27	69210	69210	100.00
	Total	307.42	100.00	224.83	100.00	163820	125936	76.87

N.B.: Small part of Mahanadi basin in Maharashtra has not to think cover its geographical area is 90 square km.
Source: Government of Maharashtra, Second Irrigation Commission Vol. July 1999, p. 205.

meters. Out of this, utilization of water for different purposes is found to be 1,03,028 million cubic meters (74 per cent of available water). Thus, utilization for irrigation purposes and for other purposes were found to be 87 per cent and 13 per cent respectively in this year. Therefore above facts indicate that utilized water for different purposes in the year 2030 A.D. would be considerably higher than utilized water in the year 1996.

Table 2
Utilization of water for different purposes in the year 1996 and 2030 A.D.

(Water in Million cubic meters)

Sr. No.	*Purpose of utilization*	*Year 1996*	*Percentage to total utilization*	*Percentage to total availability*	*Year 2030*	*Percentage to total utilization*	*Percentage total to availability*
1	2	3	4	5	6	7	8
1.	Drinking	3553.56	9	2.5	7212	7	5
2.	Irrigation						
	(a) Surface	24085.24	61	17.3	69.28	67	49
	(b) Ground	7107.12	18	5.1	20608	20	15
	Total (A+B)	31192.36	79	22.4	89634	87	64
3.	Industry	4738.08	12	3.4		6	5
4.	Total	39484	100.0	28.3	103028	100.00	74
5.	Total availability	139227		100.00	139227		100.00

Source: Government of Maharashtra, Water and Irrigation Commission, Vol. I. July 1999, p. 795.

THE ULTIMATE IRRIGATION POTENTIAL

Second Irrigation Commission of Maharashtra State was recently established in 1995 and this Commission submitted its report to Maharashtra Government in October 1999. This Commission has estimated the ultimate irrigation potential of the state to be 9.7 mha., out of which 4.1 mha. to be from the ground water resources (based on fifth water land valuation) and 5.6 mha. to be from the surface water resources.

Considering the availability of water resources and land in various basins, this commission expected that, increase in ground water land, addition of water shed development, modern irrigation technology and improvement in water management distribution on land, the ultimate irrigation potential of the state would increase up to 12.6 mha. It means due to all above factors there is possible increase in surface water resources from 5.6 mha. to 8.5 mha. Thus, the ultimate irrigation potential of the state would be 12.6 mha. of which 8.5 mha. (67.46 per cent of total potential) and 4.1 mha. (32.54 per cent) would be from surface and ground water resources respectively.

PLAN-WISE OUTLAY ON IRRIGATION PROJECTS AND POTENTIAL CREATED

Before the formation of Maharashtra state (1960). the position of irrigation in the state was very poor. After the First Five Year Plan 1951-96, Government of Maharashtra tried to develop many irrigation schemes. In the state efforts are being made to develop irrigation facilities through large, medium and major irrigation projects. Prior to inception of five year plans, the irrigation facilities to an area of only 2.74 lakh ha. was available i.e. 2.52 lakh ha. from major and medium and 0.22 lakh ha. from minor projects and an amount of Rs. 16.60 crore only was spent for this purpose. However, irrigation development programmes got real boost on inception of the five year plans, particularly from second plan onwards.

The plan wise outlay in major, medium and minor projects and created irrigation potential in Maharashtra in shown in the table 3. It is seen from this table that, in each plan period, provision for irrigation development was around 15 to 25 per cent of the total plan outlay. Up to the end of Eighth Five Year Plan (1992 to 1997), total investment of Rs. 10,756.86 crore i.e. 22 per cent of the state's plan outlay was made on irrigation development, out of which 88.48 per cent was spent on major and medium projects and remaining 11.52 per cent on minor irrigation projects. It means less amount has been spent on minor projects as compared to major and medium projects. It has increased the irrigation potential from 3.14 lakh ha. to 32.21 lakh ha. during 1951-52 to 1996-97. It means there has been near about 10 times increase in irrigation potential during the four and half decades.

REGION WISE PRESENT STATUS OF IRRIGATION PROJECTS

For smooth functioning entire irrigation system of Maharashtra has been divided into six regions, viz. Amravati, Konkan, Marathwada, Pune, Nagpur and North Maharashtra. Region-wise present status of irrigation projects completed, on going and proposed by the end of the year 1997-98 in Maharashtra state is shown in the table 4. It is seen from this table that, at present 50 major, 182 medium and 2,880 minor irrigation projects completed by the end of the year 1997-98. There are 49 major, 128 medium and 856 minor projects (469 minor state level and 387 minor local level) are presently under construction and very soon few of them will be providing irrigation facilities in their command areas. Further, there is a scope for 26 major, 63 medium and 2,394 minor irrigation projects (1162 minor state level & 1232 local minor level) to be undertaken in future.

Since Government of Maharashtra has established various Irrigation Development Corporations in different regions of the state i.e. Krishna Valley Development Corporation, Vidarba Irrigation Development

Table 3

Plan -wise Outlay on Irrigation Development & potential created in Maharashtra State

(i) Outlay in Rs. Crores
(ii) Area in lakh ha.

Sr. No.	*Plan period*	*Minor*	*Major or medium*	*Total*	*Total plan expenditure in all sectors*	*Percentage expenditure in irrigation to total*	*Cumulative irrigation potential created*	*Irrigation potential in plan period*
1	2	3	4	5	6	7	8	9
1.	Pre-plan (Before 1951)	NA	NA	16.60	–	–	2.74	2.74
2.	I (1951-56)	2.52	5.59	8.11	154.71	5.24	3.14	0.40
3.	II (1956-61)	2.74	31.20	33.94	226.30	15.00	3.98	0.84
4.	III (1961-66)	6.77	58.22	64.99	434.73	15.00	5.70	1.72
5.	Annual Plans (1966-69)	12.87	46.92	59.79	386.00	15.54	7.18	1.48
6.	IV (1969-74)	47.80	185.27	233.07	1004.51	23.20	11.02	3.84
7.	V (1974-78)	51.40	374.11	425.51	2660.00	16.00	15.70	4.68
8.	Annual Plans (1978-80)	25.62	307.53	333.15	1185.08	28.10	17.21	1.50
9.	VI (1980-85)	164.65	1138.66	1303.31	6549.39	20.00	22.70	5.50
10.	VII (1985-90)	241.24	2772.47	3013.71	11219.64	26.86	26.31	3.61
11.	Annual Plans (1990-92)	65.78	995.51	1061.29	5550.00	19.12	27.17	0.86
12.	VIII (1992-97)	607.21	3602.68	4209.99	18472.00	22.70	21.21	5.04
	Total (1991-97)	1238.70	9518.16	10756.86	47842.36	22.48		

N.B: = Not available

Source: (i) Report of High Power Committee, Irrigation Department, Government of Maharashtra , November 1981, p.86.
(ii) Performance Budget, Irrigation Department, Vol. 1, Government of Maharashtra , Vol. I, (1995-96 and 1996-97).
(iii) Second Irrigation Commission, Government of Maharashtra , Vol. I July 1999, p. 614.

Corporation, Konkan Irrigation Development Corporation, Tapi Irrigation Development Corporation and Godavari Marathwada Irrigation Development Corporation, a number of major, medium, and minor irrigation projects have been taken by the state Government through these corporations. Thus, the total irrigation potential created in the state by the end of January 1999 through all the irrigation projects was 35.6 lakh hectares (provisional)[6].

Table 4
Region-wise Present Status of Irrigation Projects Completed, On-going and future in Maharashtra

Sr. No.	*Region*	*Major irrigation project*			*Medium irrigation projects*		
		Completed	*On going*	*Proposed*	*Completed*	*On going*	*Proposed*
1	2	3	4	5	6	7	8
1.	Amaravati	6	7	2	31	22	3
2.	Konkan	3	4	3	3	6	18
3.	Marathwada	4	8	-	63	32	1
4.	Pune	14	16	7	21	26	11
5.	Nagpur	9	6	11	34	11	24
6.	North Maharashtra	14	8	3	30	31	6
	Maharashtra State	50	49	26	182	128	63

Table 4 (Continued)

Sr. No.	*Region*	*Minor irrigation projets*								
		Completed			*On going*			*Proposed*		
		State Level	*Local Level*	*Total*	*State Level*	*Local Level*	*Total*	*State Level*	*Local Level*	*Total*
		9	10	11	12	13	14	15	16	17
1.	Amaravati	144	301	445	63	39	102	186	124	310
2.	Konkan	66	111	177	30	64	94	59	229	288
3.	Marathwada	372	478	850	105	88	193	100	419	519
4.	Pune	332	105	437	119	11	130	435	290	725
5.	Nagpur	156	634	790	46	159	205	183	49	232
6.	North Maharashtra	99	82	181	106	26	132	199	121	320
	Maharashtra State	1169	1711	2880	469	387	856	1162	1232	2394

Source: Model-Maharashtra Irrigation Development-Present Position in 1996, Working Committee of Second Maharashtra State Irrigation Commission, Aurangabad, p. 145.

It is also seen from the graphs [1(A), 1 (B) and 1 (C)] that the total 99 major, 310 medium and 6,130 minor irrigation projects were undertaken out of which 50 major (50 per cent), 182 medium (58.70 per cent) and 2,880 minor irrigation projects (46.98 per cent) are already completed by the end of the year 1998. Similarly, 49 major, 128 medium and 856 minor projects are presently under construction and remaining 26 major, 63 medium and 2,394 minor irrigation projects to be undertaken in future.

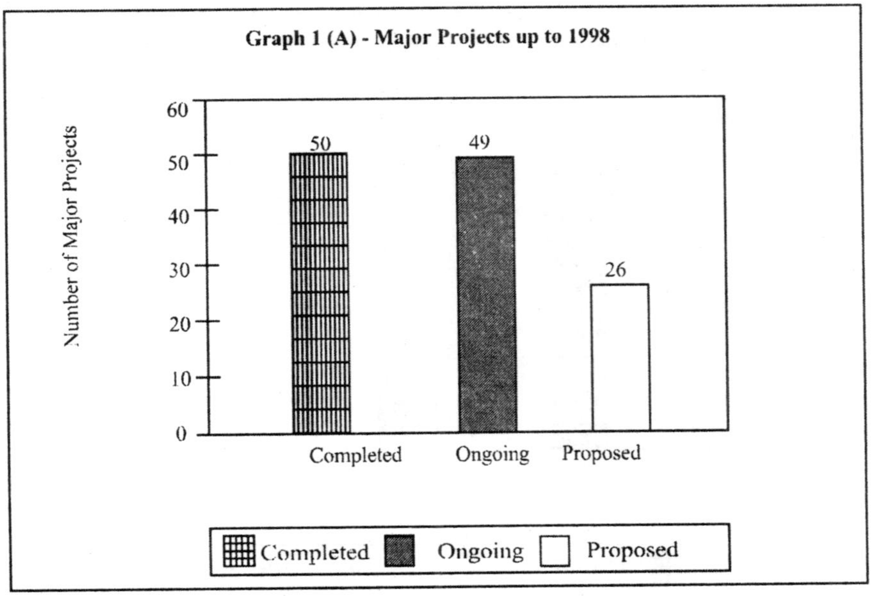

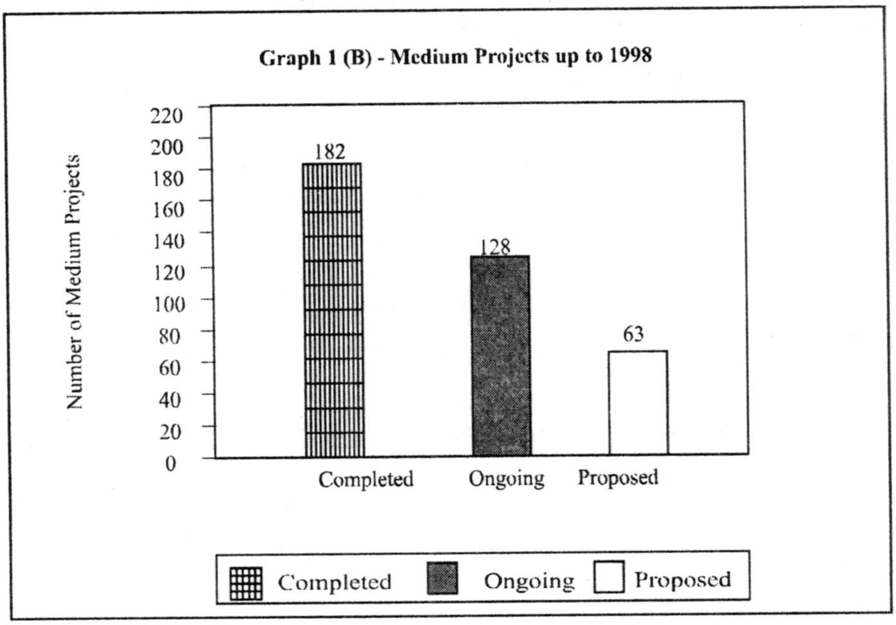

DEVELOPMENT OF LARGE DAMS IN MAHARASHTRA

From the 19th century, quantitative progress of large dams of the state and whole country has been shown in table 5. It is seen from this table that, total of 3,596 large dams were completed and 695 are under construction in the country during the period 1900 onwards, out of which

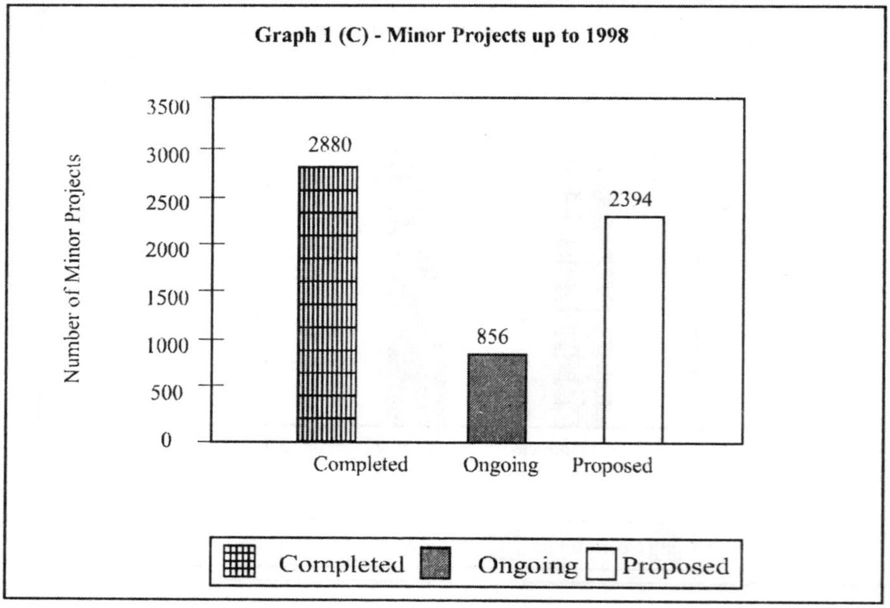

1,229 (34.18 per cent of total completed large dams) were found to be only in Maharashtra state. It means highest number of large dams were completed in the state of Maharashtra compared to other states in the country. After completing all proposed projects, live storage capacity of reservoirs in the state would be near about 54.19 cubic kilometers It would be 15.8 per cent of total live storage capacity of reservoirs in the country.[7] We can say that Maharashtra state is at the top in case of creation of storage capacity of water. However, the percentage of cropped irrigated area is very small i.e. 15 per cent which is 35 per cent at the country level.

Table 5
Quantitative Progress of Large Dams in Maharashtra State and India

Sr. No.	*State / India*	*No. of dams completed during the period*							*Under-taken*	*Total*
		Up to 1900	*1901 to 1950*	*1951 to 1960*	*1961 to 1970*	*1971 to 1980*	*1981 to 1989*	*1990 and onwards*		
1	2	3	4	5	6	7	8	9	10	11
1.	Maharashtra	16	35	25	146	589	324	94	300	1529
2.	India	42	251	234	461	1190	1066	352	695	4291

Source: 1. Water And Related Statistics, Central Water Commission , New Delhi, July, 1995, p. 50
2. Dam Safety Directorate , Central Water Commission.

Picture of quantitative progress of the large dams in India and Maharashtra during the twentieth century has been shown in the graph

2. It is seen from this graph that, during the twentieth century, there were total 4,291 dams undertaken and out of which 3,596 (83.80 per cent) dams are already completed and 695 (16.20 per cent) are found to be under construction in India. In case of Maharashtra, 1,529 dams have been undertaken and out of which 1,296 (84.76 per cent) are already completed and 300 (15.24 per cent) are found to be under construction during the same period. This shows that the Maharashtra state is well ahead (33.90 per cent) in completion of large dams as compared to other states in India.

Graph 2 - Progress of Large and Small Dams in India and Maharashtra during 20th country

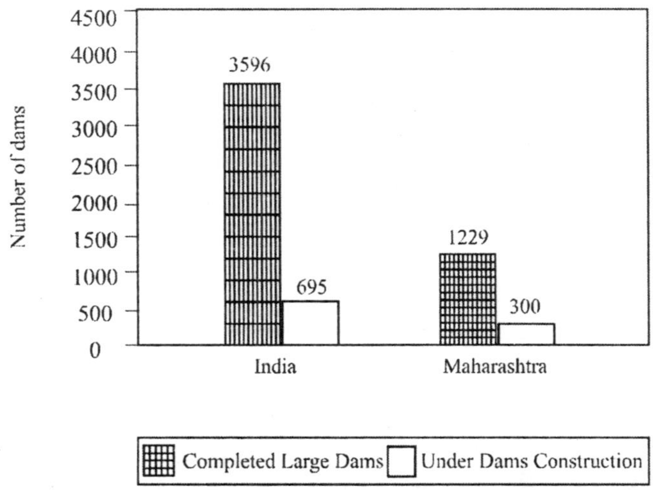

DEVELOPMENT OF IRRIGATION POTENTIAL AND UTILIZATION

The plan wise irrigation potential created and utilized from major, medium and minor projects is shown in table 6. It is seen from this table that prior to the inception of five year plans, irrigation facilities were created on an area of only 2.74 lakh ha (i.e. 2.52 lakh ha. from major and medium projects and 0.22 lakh ha. from minor projects) and its utilization was 72.27 per cent of total created potential. It means position of irrigation development in Maharashtra has given more priorities to irrigation development from beginning of the planning, particularly from the second plan onwards. It is seen from this table that irrigation potential created from all sources in the first five year plan (1951-56) was 3.14 lakh ha. Out of this, 2.75 lakh ha. (87.58 per cent) was from major and medium projects and remaining 0.39 lakh ha. (i.e.12.42 per cent) from minor projects. During 1951-52 to 1996-97 irrigation potential created rose from 3.14 lakh ha. to 32.28 lakh

Table 6
Plan-wise Irrigation Potential Created and Utilization in Maharashtra (Cumulative)

(Area in lakh ha.)

Sr. No.	*Plan period*	*Major/Medium*		*Minor state*		*Minor lift*		*All Sources*		*Percentage of utilization*
		Potential	*Utilization*	*Potential*	*Utilization*	*Potential*	*Utilization*	*Potential*	*Utilization*	
1	2	3	4	5	6	7	8	9	10	11
1.	Pre-plan (Before1951)	2.52	1.88	0.22	0.10	–	–	2.74	1.98	72.27
2.	I Plan (1951-56)	2.75	2.04	0.39	0.22	–	–	3.14	2.26	71.97
3.	II Plan (1956-61)	3.23	2.34	0.75	0.4	–	–	3.98	2.76	69.35
4.	III Plan (1961-66)	4.51	3.20	1.19	0.55	–	–	5.70	3.75	65.79
5.	Annual Plans (1966-69)	5.71	3.23	1.47	0.82	–	–	7.18	4.05	56.41
6.	IV Plan (1969-74)	8.37	4.10	2.45	0.82	0.20	0.06	11.02	4.98	45.19
7.	V Plan (1974-78)	11.24	5.73	3.38	1.13	1.08	0.10	15.70	6.96	44.33
8.	Annual Plans (1978-80)	12.32	5.77	3.81	1.30	1.08	0.12	17.21	7.19	41.24
9.	VI Plan (1980-85)	16.80	6.56	4.78	1.23	1.12	0.17	22.70	7.96	35.07
10.	VII Plan (1985-90)	19.44	8.10	5.75	1.46	1.12	0.18	26.31	9.82	37.32
11.	Annual Plans (1990-92)	19.95	8.12	6.10	1.48	1.12	–	27.17	9.60	35.33
12.	Eighth Plan (1992-978) (Target)	24.41	–	7.34	–	–	–	31.75	–	–
13.	Annual Plan (1992-93)	20.41	9.64	6.37	1.36	1.12	–	27.07	11.00	39.32

Contd....

Sr. No.	*Plan period*	*Major / Medium*		*Minor state*		*Minor lift*		*All Sources*		*Percentage of utilization*
		Potential	*Utilization*	*Potential*	*Utilization*	*Potential*	*Utilization*	*Potential*	*Utilization*	
1	2	3	4	5	6	7	8	9	10	11
14.	Annual Plan (1993-94)	12.47	10.32	7.15	1.60	–	–	29.62	11.92	40.24
15.	Annual Plans (1994-95)	23.10	–	7.41	–	–	–	30.51	–	–
16.	Annual Plan (1995-96)	–	–	–	–	–	–	–	–	–
17.	Annual Plan (1996-97)	–	–	–	–	–	–	–	–	–

N.B.: Potential Created from Minor Irrigation Including the Lift Irrigation Schemes.

Source: 1. Model :Maharashtra Irrigation Department-Present Position, working committee of second Maharashtra Irrigation Commission, Aurangabad 1996, p. 145.

2. Second Maharashtra Commission, Government of Maharashtra, Vol. I, 1999, p. 614.

hectares. It means there has been near about 11 times increase in irrigation potential created during last 46 years in the state. At the end of the annual plan of 1994-95, irrigation potential created was 30.51 lakh ha. out of which, 23.10 lakh ha. (i.e. 75.71 per cent) was created from major and medium projects and 7.41 lakh ha. (24.29 per cent) from minor irrigation projects . During the period from 1951-52 to 1994-95, irrigation potential from major and medium projects rose from 2.75 lakh ha. to 23.10 lakh ha. It means irrigation potential from major and medium projects has increased by seven times during last 44 years. In case of minor irrigation potential it rose from 0.39 lakh ha. to 7.41 lakh ha. and the increase was around 20 times in the same period. Thus, during last 46 years, more irrigation facilities were created from minor projects compared to major and medium projects in the Maharashtra state.

According to second Irrigation Commission (1995), ultimate irrigation potential of the state would be 126 lakh ha. Only of an area of 33 lakh ha. (26.19 per cent) has been already created irrigation potential in the state in the year 1997 and remaining irrigation potential of an area of 93 lakh ha. (73.81 per cent) should be created in future period of 33 years.

The percentage of utilized irrigation potential from all projects has also been shown in table 6. In case of Maharashtra, the position of utilization of potential created is very poor. It is seen from this table that during the period 1951 to 1966, the percentage of utilized irrigation potential was 65 to 70 per cent but after this period it has been continuously decreased. At the end of the annual plans (1990-92), it was only 35 per cent of created potential. At the end of annual plan of 1993-94, the per centage of utilized irrigation potential was found to be 40 per cent.

Irrigation Potential-Ultimate, Created and Utilized

State's ultimate irrigation potential, created potential and its utilization in case of major, medium and minor projects up to the period 1997 are presented in graph 3. It is seen from this graph that the ultimate irrigation potential has been recently estimated for state to an area of 126 lakh ha. Out of this 32 lakh ha.(25.39 per cent of ultimate potential) is potential created from all irrigation projects and only 10 lakh ha. (31.25 per cent of created potential) is potential utilized by the end of 1997.

SOURCE-WISE IRRIGATION DEVELOPMENT IN MAHARASHTRA (1960-61 TO 1999-2000)

The data about net irrigated area, gross irrigated area, gross cropped area and percentage of gross irrigated area to gross cropped area of Maharashtra (1996-61 to 1999-2000) is presented in table 7. It is seen

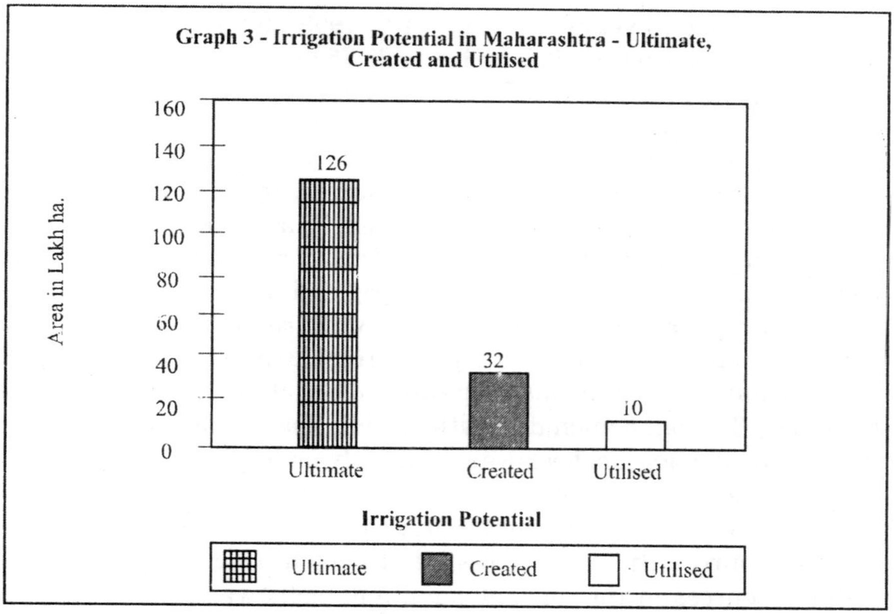

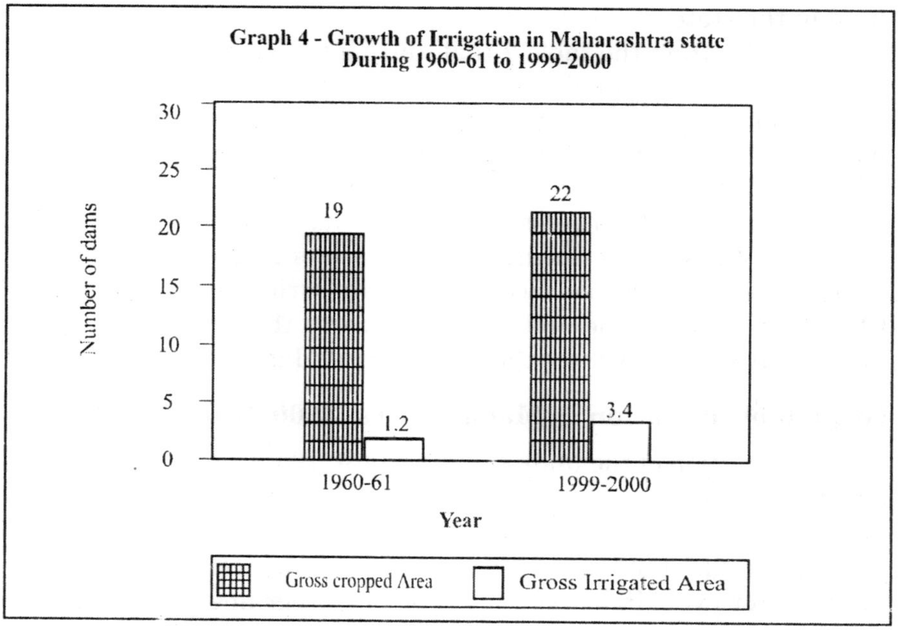

from this table that, net irrigated area was 10.72 lakh ha. in the year 1960-61 from surface and ground water resources, out of which, 5.95 lakh ha. (55.50 per cent of net irrigated) from the surface water. It means that more than 50 per cent area was irrigated from wells, which was found to be more dominant source of irrigation in the state.

During last 40 years (i.e. 1960-61 to 1999-2000), net irrigated area in the state rose from 10.72 lakh ha. to 25.75 lakh ha. Thus, net irrigated area of the state has increased by 140.20 per cent. Out of the net irrigated area of 25.75 lakh ha.,14.07 lakh ha. (56.64 per cent of total net area) is from well irrigation and 11.68 lakh ha. (43.36 per cent of total net irrigated area) is from the surface resources. It means during last 40 years, there was 135.68 per cent increase in irrigated area from wells and 144.86 per cent from other sources. Table 7 also indicates that during last 40 years there was increase in irrigated area under surface and ground water resources. Therefore, Maharashtra state has more potential in both surface and well resources during last forty years. But well irrigation source has remained more dominant than surface resources. The phenomenal growth of wells shows the farmers choice of this source of irrigation because it is directly under the control of farmers and it provides water sufficiently and timely to control of farmers and it provides water sufficiently and timely to cultivated crops. While in case of other sources this is not possible. Hence during last four decades, water resources in the form of canal and wells were being utilized in various regions and the irrigation facilities are found to be intensively used in the state.

It is seen from the table 7 that, during the period 1960-61 to 1999-2000, the gross irrigated area of the state rose from 12.20 lakh ha. to 33.74 lakh ha. In other words, gross irrigated area of the state has increased by 176.56 per cent. Similarly, gross cropped area rose from 188.23 lakh hectares to 218.97 lakh hectares. Therefore, percentage of gross irrigation area to gross cropped area have increased from 6.48 per cent to 15.41 per cent during last forty years in the state. There was sharp increase in well, surface, net and total irrigated area during 1960-61 to 1999-2000 in the state. This indicates that, irrigation facilities intensively used only after the formation of Maharashtra state(1960).

Growth of Irrigation in Maharashtra (1960-61 to 1999-2000)

Before formation of the Maharashtra State (1960), position of irrigation development was very poor. The gross irrigated area of the state was amounted to a mere 6 per cent of gross cropped area and the irrigation potential in the state amounted to a mere 2.16 per cent of net sown area in 1960.However, irrigation development programs got real boost on inception of the five year plans, particularly from second plan onwards. Thus, with investment in irrigation schemes, it become possible to undertake 125 major, 373 medium and 8130 minor projects during1960-61 to 1999-2000. The Government of Maharashtra has been provided large amount of money for irrigation development schemes through successive plans .Thus, the state's irrigation potential has increased from 3.14 lakh hectares to 32.21 lakh hectares .

Table 7
Net and Gross Cropped Area in Maharashtra State. (1960-61 to 1999-2000)

(Area in lakh ha.)

Sr. No.	Year	Area Irrigated				Gross Cropped area	% of gross irrigated area to gross cropped area
		Wells	other sources	Total Net area Irrigated	Gross irrigated area		
1	2	3	4	5	6	7	8
1.	1960-61	5.95 (55.50)	4.77 (44.50)	10.72 (100)	12.20 (100.00)	188.23	6.48
2.	1961-66	7.11	4.96	12.06	13.88	189.72	7.32
3.	1970-71	7.68	5.79	13.47	15.70	187.37	8.38
4.	1975-76	10.84	7.17	18.02	21.71	196.64	11.04
5.	1980-81	10.55	7.80	18.35	24.15	196.42	12.30
6.	1985-86	11.62	7.86	19.49	24.20	205.69	11.77
7.	1988-89	14.19	9.71	23.90	30.74	214.86	14.31
8.	1989-90	15.68	10.07	25.70	32.17	216.62	14.85
9.	1990-91	17.72	9.99	26.71	33.19	218.59	15.18
10.	1991-92	10.32	9.81	27.13	32.52	201.33	16.15
11.	1992-93	17.37	9.43	26.80	32.61	211.89	15.39
12.	1993-94	13.73	11.41	25.14	32.93	214.09	15.38
13.	1994-95	13.77	11.44	25.21	33.01	241.18	15.41
14.	1995-96	13.71	11.39	25.10	32.87	213.27	15.41
15.	1996-97	13.92 (54.61)	11.57 (45.39)	25.49 (100)	33.38	216.62	15.41
16.	1997-98	13.97	11.61	25.59	33.50	21.40	15.41
17.	1998-99	14.02	11.62	25.68	33.62	218.18	15.41
18.	1999-00	14.56 (56.64)	11.68 (43.36)	25.75	33.74	218.97	15.41
		135.68	144.86	140.20	176.56	16.33	–

N.B.: Figures in brackets indicate the percentage to total net irrigated area.
Source: Epitome of Agriculture in Maharashtra State, Pune, pp. 28, 31.

The data about gross irrigated area and gross cropped area in Maharashtra during 1960-61 to 1999-2000 is shown in the graph 4. It is seen from this graph that just about an equivalent of 1.2 million ha. of irrigation benefits in the year 1960-61, the corresponding benefits near about tripled by the year 1999-2000 to reach 3.4 million ha. in Maharashtra. As compared to just about 6 per cent of the cropped area under irrigation in the period 1960-61, the irrigation level in state has since increased to 15.41 per cent in the year 1999-2000 though there has been an increase of nearly 20 per cent in the cropped area.

Thus, the quantitative aspects of irrigation facilities created in the state appear more depressing when viewed against the total available irrigation potential of the state from all sources. Out of the total potential estimate to irrigate a net area of 89 lakh hectares or 34.28 per cent has been created up to 1994-95 and only 40 per cent has been utilized. This inadequacy of irrigation in Maharashtra appears less striking once we compare it with the level of irrigation development in India. In 1995-96, the percentage of gross irrigated area to gross cropped area was 15.41 per cent in Maharashtra state, which was less than the corresponding all India average of 38.00 per cent.

NOTES AND REFERENCES

1. Bhagirath, vol xxxx vi, April-June 1999, Central water commission and ministry of water resources, New Delhi, pp. 78-82.
2. Arunachalam, B.,'Maharashtra', published by Mahendrakumar A Sheth for A. K. Sheth and co., Princess Street, Bombay 1967, pp. 3-4.
3. Economic Survey of Maharashtra, 1999-2000 Directorate of Economics and Statistics, Planning Department , Government of Maharashtra, Mumbai, p. T 31.
4. Report of the Irrigation Commission, 1972, Vol. II, Ministry of Irrigation and Power, New Delhi, p. 237.
5. Report of the Fact Finding Committee on Regional Imbalance, Government of Maharashtra, 1984, p. 84.
6. Economic Survey of Maharashtra, 1999-2000, Directorate of Economics and Statistics , Planning Department, Government of Maharashtra, Mumbai.
7. Central Water Commission, 1996, Government of India, Statement No. 18, p. 17.

A.N. Pawar

IMPACT OF IRRIGATION ON AGRICULTURE

A Macro-Level Analysis

6

Irrigation has played a vital role in continuous process of agricultural development in India. In successive five year plans irrigation development has been given top priority. Thus the irrigation facility in India rose from 22.60 million hectares to 91.79 million hectares during last 46 years(ie 1951-52 to 1995-96). As compared to just about 18 per cent of cropped area under irrigation at the time of independence, the irrigation level has since increased to 38 per cent in 1995-96 though there has seen an increase of nearly 42 per cent in cropped area.

At the time of commencement of the first five year plan in 1951, population of India was about 361 million and annual food grain production was 51 million, which was not adquate. Import of food grains was then inevitable to cover up the shortage. Attaining self-sufficiency in food was therefore given paramount importance in the plan period and in order to achieve the objective, various irrigation projects were formulated and implemented though successive five year plans. This drive compounded with Green Revolution in the agricultural sector has enabled India to become marginally surplus from surplus from a deficit one in food grains. Thus, food production has touched nearly 200 million tonnes mark in the year 1996-97.[1]

Before the formation of Maharashtra state, position of irrigation development in the state was very poor. However, irrigation development programe since got real boost inception of five year plans. Thus, irrigation potential in the state has increased from 3.14 lakh hectares to 32.21 lakh hectares during 1951-56 to 1992-97. The additional irrigation facility has been provided by the completed 50 major 182 medium and 2280 minor projects and partial on going 49 major, 128 medium and 856 minor projects in the state. Thus, it was just about 6 per cent of cropped area was under irrigation at the time of formation of the Maharashtra state (1960). The irrigation has increased in the state to

15.40 per cent in the year 1998-99 though. There has been an increase to nearly 20 per cent in copped area.

Before the formation of the Maharashtra state in 1960, it is observed that the irrigation development in Marathwada region was negligible. In 1950-51 there was only one medium project in Beed district. After the formation of state, Government has given top priority to development of irrigation schemes in the Marathwada region. The irrigation potential has increased from 11.64 thousand hectares to 8.97 lakh hectares. During 1960-61 to 1998-99 4 major, 63 medium & 3112 minor irrigation projects have been already completed. In addition 7 major, 19 medium and 76 minor projects are on the way of completion. Thus, after exploiting all irrigation projects, it would be possible to bring under irrigation 25.30 per cent of cultivated area in Marathwada region.

Irrigation plays an important role in agricultural growth and development. Therefore it is necessary to insure the irrigation development create favorable impact on agriculture and ultimately on the life of the rural households. From this point of view, it is necessary to assess the irrigation impact on agriculture development. The study is under taken with this view. The district like Osmanabad (in Marathwada region) in Maharashtra State is selected for the present study. Agriculture is the main occupation of the people in this district and more than 80 per cent of workers depend on agriculture sector for their livelihood now 17 Per cent of gross cropped area is under irrigation in the district.

This study was intended to following objectives to evaluate the major impacts of irrigation on Agriculture in Osmanabad district.

(i) To review the development of irrigation in Osmanabad district.

(ii) To assess the impact of irrigation on land use pattern in Osmanabad district.

(iii) To assess the impact of irrigation on cropping pattern in Osmanabad district.

(iv) To assess the impact of irrigation on crop production in the study area

(v) To assess the impact of irrigation on cropping intensity in Osmanabad district.

RESEARCH METHODOLOGY

An attempt is made in this paper to examine macro-level impact of irrigation on agriculture in Osmanabad district. Secondary data was collected from various publications of state and local government offices. The study was based on before-after approach. The period for macro-level study is considered from 1982, since taluka level published data were available for Osmanabad district. Before 1982, the Latur taluka was part of Osmanabad district which was formed as a new district in

1982. So the taluka-wise data for Osmanabad district has been taken from 1982 onwards.

Statistical Methods: The study is attempted to make use of the following statistical methods.

(a) The intensity of irrigation and cropping both were computed by following formula.

(i) Irrigation Intensity Index

Gross irrigated area
Net irrigated area

(ii) Cropping Intensity Index

Gross cropped area
Net cropped area

(b) Correlation coefficient were calculated between total irrigated area and total crop production irrigated area under food grains and food grain production and irrigation area under non food grains and non food-grains production with the following formula:

Correlation coefficient or

$$r = \frac{\sum xy - N\overline{X}\,\overline{Y}}{N\sigma\chi.\sigma\bar{y}}$$

Where X = Actual mean of X values

X = Total irrigated area/irrigated area under food grains/ irrigated area under non-foodgrains

Y = Actual mean of Y values

Y = Total crop production /food grains production/non-foodgrain production

σx = standard deviation of X values

σy = standard deviation of Y values

MACRO LEVEL IMPACT OF IRRIGATION ON THE FARM ECONOMY

Impact of irrigation development on a less developed farms economy can be reaching both in dimension and magnitude. In Osmanabad district, in terms of investment considerable importance seems to have been attached to irrigation during 1982-83 to 1995-96. Therefore, in this study, the various effects of irrigation on agricultural economy of Osmanabad district have been examined.

In the precise context of agricultural production, the direct effects of irrigation can be said to consist of (a) area or acreage effect, (b) crop pattern effect. (c) per hectare yield effect, and (d) total output effect. Firstly, irrigation leads to an expansion of gross cropped area by making double and multiple cropping possible. Secondly, it raises yield per unit of area by inducing the use of other complementary yield raising inputs i.e. fertilizers, pesticides, high yielding variety seeds, etc. Thirdly, for given set of output and input prices, irrigation may raise production by enabling farmers to allocate their lands to high yielding and high valued crops. Lastly, irrigation may raise production by highly utilizing the land resources. Investment in irrigation becomes an indirect investment for eradicating much of the rural poverty. Therefore, the development of irrigation is found to be emphasized, with irrigation acting as a Iynchpin in the entire effort directed towards rural development in India.

Certain important studies undertaken on this issue may briefly be summarized. Ghosh has indicated the pervasive character of irrigation impact that operates through interaction of area, yield and crop pattern phenomena.[2] Bhardwaj has shown that irrigation can raise the productivity in other ways i.e. by making possible multiple cropping, by increasing the yield per unit cost and allowing the production of more lucrative crops.[3]

Gadgil concluded that the main direct effect of irrigation is to put the previously cultivated lands to new uses or to make them more productive in former uses to bring new lands under use for the first time.[4]Mukhopadhyaya illustrated this through the positive and statistically significant co-effcient of interaction between irrigation and fertilizers.[5] Singh et al., as well as Gaikwad and Desai have shown that the provision of irrigation causes higher crops yields per unit of land and generates increased agricultural incomes.[6] Similarly, the number of scholars from various regions of India have all confirmed the productive role of irrigation while highlighting its contribution for higher rates of agricultural growth. Now each of these effects as they operate in Osmanabad district are discussed in the present paper.

Thus an attempt has been made in the present paper to assess the impact of irrigation on agriculture in Osmanabad district. In this paper an attempt has been made to assess the macro level impact of irrigation on agriculture in all talukas of Osmanabad district and the district as a whole during 1982-83 to 1995-96. This paper has been divided into five parts. The first part deals with the impact of irrigation on cultivated area i.e. growth in net, total, and double cropped area. The second part deals with the impact of irrigation on intensity of cropping. In the third part and attempt has been made to study the impact of irrigation on cropping pattern. In the fourth part an attempt has been made to study impact of irrigation on yield and in the fifth part, an attempt has been made to study the correlation between irrigation and total crop production. The statistical techniques used in this paper are already discussed above.

AREA EFFECT OF IRRIGATION

An impact of irrigation through which it increases the gross cropped area is called its "area effect" or "cultivated area effect". This phenomenon consists of three components: first increase in net sown area which becomes possible on account of irrigation facilitating reclamation of waste land and extending cultivation to hitherto uncultivated land, second increase in double cropped area which is directly related to the irrigation facilities in any region. Thus double cropping is possible only with the development of irrigation facilities and shows how many times a piece of land is used for crop production. Thus, lastly, it increases gross cropped area, that arises out of an irrigation encouraging the intensive cultivation of currently cultivated land, that is the index of cropping intensity. Both these components through positive interaction lead to absolute increase in gross cropped area and thereby 'area effect" or cultivated area effect".

Net Cropped Area

The first component of "acreage effect" operated in Punjab and in regions served by Rajasthan canals, where the cultivated areas increased significantly on account of reclamation of waste land and progressive decline in fallow land, consequent upon the availability of irrigation water[7] but in district like Osmanabad of the Maharashtra state this cultivated area does not seem to have increased much long with the extension of irrigation facility. The total geographical area, net cropped area, double cropped area and total cropped area for six talukas in the Osmanabad district and district as whole during 1982-83 to 1995-96 has been shown in table 1. And percentage share of net cropped area to total geographical area has been calculated for six talukas in the district and for the district as a whole during 1982-83 to 1995-96and is presented in the table 2. It is seen from the tables 1 and 2 that, during last fourteen years (1982-83 to 1995-96), there was increase of 4.18 per cent in net cropped area in Osmanabad district. This shows that, there was positive increase in irrigation and net cultivated area of the district.

It is also seen from the table 1 that, in 1982-83 the irrigated area in Osmanabad district was 66,700 hectares and the net cultivated area was 5,61,200 hectares. But in 1995-96 while irrigated area went up to 1,00,000 hectares i.e .it was increased by 49.92 per cent, the cultivated area went up to 5,92,500 hectares, i.e. an increase of 5.28 per cent. Thus the growth of irrigation in Osmanabad district has, therefore, little impact on the first component of area effect.

A perusal of table 2 shows, that during the last fourteen years (1982-83 to 1995-96) there was a gradual increase in net cultivated area. During last fourteen years, the percentage share of net cropped area to total geographical area in the district has registered an increase of 4.18 per

Table 1
Taluka wise Total Geographical Area and Cultivated Cropped Area in Osmanabad District(1982-83 to 1995-96)

(Area in hundred hectares)

Taluka / District	*Year*	*Total geographical area*	*Net cropped area*	*Area cropped more than once*	*Total cropped area*	*Net irrigated area*
Osmanabad	1982-83	1326	1036	43	1079	188
Kallam		1227	948	30	978	78
Omerge		1465	1286	44	1330	188
Tuljapur		1524	800	40	840	188
Paranda		1055	894	51	945	103
Bhoom		887	648	21	1079	188
District		7484	5612	229	5841	667
Osmanabad	1986-87	1326	1093	69	1162	194
Kallam		1227	895	80	975	124
Omerga		1465	1234	269	1503	120
Tuljapur		1524	931	206	1137	228
Paranda		1055	926	169	1095	110
Bhoom		887	617	62	679	86
District		7484	5696	855	6551	862
Osmanabad	1990-91	1326	1094	143	1237	216
Kallam		1227	979	146	1125	148
Omerga		1465	1208	333	1541	176
Tuljapur		1524	1073	407	1480	356
Paranda		1055	963	291	1254	120
Bhoom		887	665	99	764	120
District		7484	5982	1419	7401	1136
Percentage change		-	5.28	544.98	26.72	49.92

Source: Socio-Economic Review and District Statistical Abstract of Osmanabad District, Directorate of Economics and Statistics, Government of Maharashtra, 1985-86, 1989-90,1990-91 and 1998-99.

Table 2
Percentage Share of Net Cropped Area to Total Geographical Area in Osmanabad District (1982-83 to 1995-96)

(Area in hundred hectares)

Taluka	*1882-83*	*1986-87*	*1990-91*	*1995-96*	*Change in Percentage (1982-83 to 1995-96)*
Osmanabad	78.12	82.42	82.50	84.09	(5.97)
Kallam	77.26	72.94	79.79	59.25	(-18.01)
Omerga	87.78	84.23	82.46	86.62	(-1.16)
Tuljapur	52.49	61.08	70.41	76.38	(23.89)
Paranda	84.73	87.77	91.28	90.05	(5.32)
Bhoom	73.05	69.56	74.97	78.92	(5.87)
District	74.98	76.11	79.93	79.16	(4.18)

N.B.: Figures in the brackets show change in percentage share of Net cropped Area to Total Geographical Area during (1982-83 to 1995-96).

Source: Estimated from the Data Collected from Socio-Economic Review and District Statistical Abstract of Osmanabad District, Directorate of Economics and Statistics, Government of Maharashtra, 1985-86,1989-90,1990-91 and 1998-99.

cent. The talukas of Tuljapur (23.89 per cent), Bhoom (5.87 per cent) Osmanabad (5.97 per cent) and Paranda (5.32 per cent) have registered higher percentage increase in net cultivated area than the district as a whole, and talukas of Kallam (-18.01) and Omerga (-1.16) have registered decrease in net cultivated area.

Gross Cropped Area

More important than this increase in cultivated area, is the second component of "area effect", that is, increase in gross cropped area resulting from the higher crop intensity. With the development of irrigation, gross cropped area has also increased in the study area. Percentage share of gross cropped area to total geographical area (from 1982-83 to 1995-96) has been calculated for Osmanabad district and for all talukas and is presented in table 3. It is seen form this table that, during the last fourteen years (1982-83 to 1995-96), the percentage share of gross cropped area to total geographical area in the study area has registered an increase of 20.85 per cent.

Table 3
Percentage Share of Gross Cropped Area to Total Geographical Area in Osmanabad District (1982-83 to 1995-96)

(Area in hundred hectares)

Taluka	*1882-83*	*1986-87*	*1990-91*	*1995-96*	*Change in Percentage (1982-83 to 1995-96)*
Osmanabad	81.37	87.69	93.29	93.29	(11.92)
Kallam	79.70	79.46	91.69	80.03	(0.33)
Omerga	90.10	102.59	105.19	102.59	(12.49)
Tuljapur	55.12	74.61	97.11	96.85	(41.73)
Paranda	89.57	103.79	118.86	118.29	(28.72)
Bhoom	75.42	76.55	86.13	107.77	(32.35)
District	78.04	87.53	98.89	98.89	(20.85)

N.B: Figures in the brackets show change in percentage share of Gross Cropped Area to Total Geographical Area during 1982-83 to 1995-96.

Source: Estimated from the Data Collected from Socio- Economic Review and District Statistical Abstract of Osmanabad District, Directorate of Economics and Statistics, Government of Maharashtra, 1985-86, 1989-90,1990-91 and 1998-99.

Among the talukas in the district, Tuljapur (41.73 per cent), Bhoom (32.35 per cent) and Paranda (28.72 per cent) talukas recorded higher increase in gross cropped area than the district average (28.85 per cent). And rest of the talukas, namely, Omerga (12.49 per cent), Osmanabad (11.98 per cent) and Kallam (0.33 per cent) have recorded lower increase in gross cropped area than the district average. Thus, the growth of irrigation in talukas of Osmanabad district and district as a whole has, therefore, positive impact on gross cropped area.

Double Cropped Area

Area under double cropping shows that how many times a piece of land is used for the crop production or how intensively the agriculture land is being utilized. Double cropping is possible only with the development of irrigation facilities. Percentage share of double-cropped area to net-cropped area (from 1982-83 to 1995-96) was calculated for Osmanabad district and for its talukas and is presented in table 4.

Table 4
Percentage Share of Double Cropped Area to Net Cropped Area in Osmanabad District (1982-83 to 1995-96)

(Area in hundred hectares)

Taluka	*1882-83*	*1986-87*	*1990-91*	*1995-96*	*Change in Percentage (1982-83 to 1995-96)*
Osmanabad	4.15	6.31	13.07	10.94	(6.79)
Kallam	3.16	8.94	14.91	35.07	(31.91)
Omerga	3.42	21.80	27.57	18.52	(15.10)
Tuljapur	5.00	22.13	37.93	26.80	(21.80)
Paranda	5.70	18.23	30.22	31.37	(25.67)
Bhoom	3.24	10.05	14.89	36.43	(33.19)
District	4.08	15.01	23.72	24.93	(20.85)

N.B: Figures in the brackets show change in percentage share of Double Cropped Area to Net Cropped Area during (1982-83 to 1995-96).

Source: Estimated from the Data Collected from Socio-Economic Review and District Statistical Abstract of Osmanabad District, Directorate of Economics and Statistics, Government of Maharashtra, 1985-86, 1989-90,1990-91 and 1998-99.

A perusal of table 4 shows that, during the last fourteen years, (1982-83 to 1995-96) there was 20.85 per cent increase in the percentage share of double cropped area to net cropped area in the study area. And during the same period the talukas, namely, Bhoom (33.19), Kallam (31.91), Paranda (25.67) and Tuljapur (21.80) recorded higher increase in the percentage share of double cropped area to net cropped area than the district average. This shows that, all the talukas in the Osmanabad district and the district as a whole recorded positive change in the percentage share of double cropped area during last fourteen years. Thus, this shows that, the growth of irrigation in Osmanabad district and almost all its talukas during the period under study have positive impact on cultivated area.

CROPPING INTENSITY INDEX

Intensity of cropping is considered as an indicator of the degree of intensity to which land is used for agricultural purpose. It is the ratio between the gross cropped area and the net cultivated area usually expressed in percentage terms. How many crops a cultivator can produce in a year on his land or in other words to what extent a cultivator will practise double cropping. Intensity of cropping will depend on several factors, such as availability of irrigation water, use of manures and the nature of crops which he decides to grow. As regards the irrigation water and manure, it may be argued that improvement in their use induce intensity of cropping. As regards the nature of crops, it may be pointed out that there are crops which can be combined and can be grown in the same agricultural year. There are certain crops like sugarcane, which are of long duration crops, do not permit another crops in the same year.[8] The fact is that irrigation alone does not constitute the sole factor determining the intensity of cropping.

Intensity of cropping is a single indicator of the extent of multiple cropping on land under plough, that is the number of crops being raised in a sequence. If a piece of land is cultivated once in a year, the index number would be 100 per cent and twice or thrice it would be more than 100 per cent.

Most important is that increase in cultivated area is the second component of "area effect", that is, increase in gross cropped area resulting from higher crop intensity. The net cropped area, gross cropped area, area cropped more than once and net irrigated area (during 1982-83 to 1995-96) for the Osmanabad district and its talukas has been shown in table 1. The intensity of cropping for the six talukas in the district and district as a whole has been calculated for the period of fourteen years (1982-83 to 1995-96) and is presented in table 5 and in graph 1. It is seen from this table and graph that throughout the period under study i.e. from 104.08 per cent it rose to 124.93 per cent in 1995-96. Thus, there was increase of 20.85 percentage of points in the intensity of cropping during the last fourteen years.

Table 5
Intensity of Cropping in Osmanabad District (1982-83 to 1995-96)

Taluka	*1882-83*	*1986-87*	*1990-91*	*1995-96*	*Change in Percentage (1982-83 to 1995-96)*
Osmanabad	104.15	106.31	113.07	110.94	(6.79)
Kallam	103.16	108.94	114.91	135.07	(31.91)
Omerga	103.42	121.80	127.57	118.52	(15.10)
Tuljapur	105.00	122.13	137.93	126.80	(21.80)
Paranda	105.70	118.25	130.22	131.37	(25.67)
Bhoom	103.24	110.05	114.89	136.43	(33.19)
District	104.08	115.05	123.72	124.93	(20.85)

N.B: Figures in the brackets show change in intensity of cropping during (1982-83 to 1995-96).

Source: Estimated on the basis of Data collected form Socio-Economic Review and District Statistical Abstract of Osmanabad District, Directorate of Economics and Statistics, Government of Maharashtra, 1985-86, 1989-90,1990-91 and 1998-99.

Graph 1 : Taluka wise Cropping Intensity in Osmanabad District During 1982-83 and 1995-96

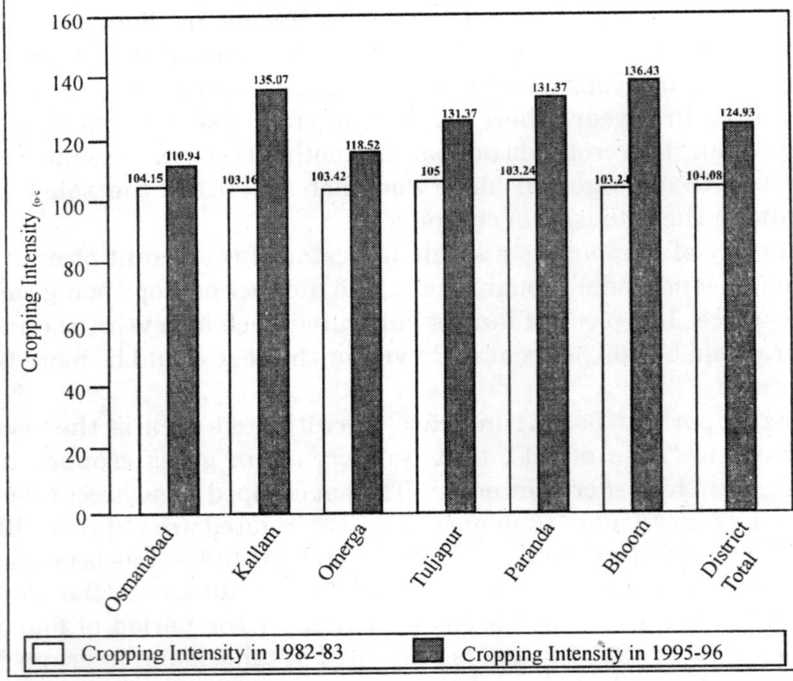

In 1982-83, highest intensity of cropping was recorded in the Paranda taluka (105.70 per cent). Tuljapur (105.70) and Osmanabad talukas (104.15) have also recorded higher percentage of cropping intensity than the district average. In these talukas irrigation facilities were highly developed. The talukas, which recorded low percentage of cropping intensity than the district average, are Kallam (103.16), Bhoom (103.24), and Omerga (103.42).

In the year 1995-96, Bhoom taluka (136.43) recorded the highest percentage of cropping intensity followed by Kallam (135.07), Paranda (131.37) and Tuljapur (126.80) talukas. They also recorded higher percentage of cropping intensity than the district average of 124.93 per cent. Thus, these figures show that, the intensity of cropping is higher in those talukas where irrigation facilities are highly developed. To some extent cropping intensity is also dependent on the cropping pattern. If a substantial proportion of the cultivated area is under long standing crops like sugarcane the intensity will be low. But the farmers in the study area find growing of sugarcane more remunerative than raising consequently two or three crops. The talukas, which recorded less percentage of cropping intensity than the district average (124.93 per cent), are Osmanabad (110.94 per cent) and Omerga (118.52 per cent).

A perusal of table 5 and graph 1 shows that during last fourteen years, there was 20.85 per cent increase in district's intensity of cropping. It is also observed that almost all the talukas have recorded a gradual increase in cropping intensity. The talukas of Bhoom (33.19 per cent), Kallam (31.19 per cent), Paranda (25.67 per cent), Tuljapur (21.80 per cent), Omerga (15.10 per cent), and Osmanabad (6.79 per cent) have recorded increase in cropping intensity. Thus the growth of irrigation in Osmanabad district and its all talukas during last fourteen years have positive impact on cropping intensity.

IMPACT OF IRRIGATION ON CROPPING PATTERN

The third aspect of irrigation impact on agriculture in the Osmanabad district is the changing crop pattern as evidenced in the district in recent years. Cropping pattern simply means the proportion of area under different crops at a point of time, whereas change in cropping pattern refers to change in proportion of area under different crops at two different points of time. Of course, all changes in cropping pattern can not be said to be irrigation induced. Some shifts in crop pattern do not take place on account of economic and institutional factors, such as urbanization, change in food habits of the people, input requirements of crops, relative price movement of the agricultural products, state policy on subsidies and change in technology. Thus, the cropping pattern of any area is generally controlled by physical, socioeconomic and technological factors. But it is often seen that irrigation is the most important factor that determines to such changes in the crop pattern and makes possible the production of more remunerative crops. For example, as Bhardwaj has pointed out that while on irrigated holdings in Punjab only 50 per cent of the cropped area was devoted to food crops, on unirrigated holdings on the other hand, 73 to 94 per cent of the cropped area went to food crops. Similarly in Tamilnadu, the irrigated holdings concentrate on paddy, a relatively more valuable crop, while the unirrigated holdings mainly produce dry grains like 'Oumbu' and 'Cholam'.[9] While classifying the effects of irrigation, Prof. D.R. Gadgil identified the change in cropping pattern as one of the direct effects of irrigation. The introduction of irrigation brings alterations in area allocations among crops. The availability of irrigation water enhances double cropped area. Farmers take more than one crop on the same piece of land and hence the crop rotation undergoes a change. Further, the proportion of area under rabi and hot weather (summer) crops also increases due to irrigation facility. Due to introduction of irrigation, the shift is observed from "low water requiring crops" to "highwater requiring crops". Before introduction of irrigation rain crops dominated cropping pattern. With the changes and availability of water crops like sugarcane, oilseeds, fruits etc., which require more water per unit of land, were encouraged. The availability of water changes cost structure because of the change in the inputs. This increases the need for earning higher profits and hence proportion of area under non-foodgrains, which gives more profit, profit increase thereby alters crop pattern. As such, it may broadly be stated that, it is the irrigation expansion that plays the dominant role in the context of crop pattern changes.

There are two important agricultural seasons. The main kharif season crops sown in Osmanabad district are paddy, jowar (hybrid), bajra, tur and oilseeds (kharif), while jowar, wheat, gram, oilseeds, fruits and vegetables are sown during the rabi season. Sugarcane is annual crop, which is considered in kharif season for this study. The crop wise cultivated area and the percentage share of crop wise cultivated area in kharif and rabi season to gross cultivated area has been calculated for the district as a whole and for its talukas for a period of fourteen years (from 1982-83 to 1995-96) and gross cropped area over a period of fourteen years has been shown in table 5.

Crop Pattern Effect of Irrigation: Kharif Season

Kharif crops are those crops which are grown during the monsoon period which is the wet season of the country. Agriculture year begins in June in India. The introduction of Kharif season irrigation in the study area has helped the farmers to switch over from low yielding dry crops to high yielding wet crops like paddy, sugarcane and oilseeds. Sugarcane cultivation with the help of perennial source of irrigation is bound to produce high output impact when introduced in low and medium rainfall areas. Kharif oilseeds are partly dependent on irrigation which is very helpful for growing oilseeds in low rainfall time of study area. Therefore, oilseeds cultivation with the help of seasonal irrigation is also found to have high output impact when introduced in low rainfall area of this district. A perusal of table 6 shows that, in 1982-83, the district's cropped area under kharif season was 2,29,887 hectares and in 1995-96 it was 3,32,2186 hectares. Thus, during 1982-83 to 1995-95, there has been 44.50 per cent increase in kharif cultivated area in the district.

A perusal of tables 6 and 7 shows that, the percentage share of area under kharif crops to the gross cropped area in the district during 1982-83 was about 39.36 per cent. In 1982, kharif jowar (16.39 per cent) occupied the highest cropped area in kharif season. Thus, kharif jowar (16.39 per cent) was the most important crop of kharif season followed by tur (9.56 per cent), oilseeds (7.63 per cent), sugarcane (2.44 per cent), paddy (2.88 per cent) and bajra (0.96 per cent). While in 1995-96, the percentage share of land under kharif crops to gross cropped area in the district was about 45.02 per cent. During last fourteen years (i.e. 1982-83 to 1995-96), percentage share of kharif cultivated area to gross cropped area in the district has increased by 5.66 per cent. In 1995-96, Oilseeds

Table 6
Cropwise Cultivated Area and the percentage Share of Area Under Different Crops to Gross Cropped Area in Osmanabad District (1982-83 to 1995-96)

(Area in hectares)

Talukas	*Year*	*Gross Cropped Area*	*Kharif Crops*							*Rabi Crops*					
			Paddy	*Jowar*	*Bajra*	*Tur*	*Sugar-cane*	*Oil-Seeds*	*Total Kharif*	*Jowar*	*Wheat*	*Gram*	*Oil-Seeds*	*Fruits & veg.*	*Total Rabi*
Osmanabad															
	1982-83	107838	2400	23297	330	9817	2031	4681	42577	30215	7785	8590	9253	305	56148
		(100.00)	(2.23)	(21.60)	(0.31)	(9.10)	(1.88)	(4.34)	(39.46)	(28.02)	(7.22)	(7.97)	(8.58)	(0.28)	(52.07)
	1986-87	116185	2310	28461	N.G.	9710	2175	11925	54631	32791	1990	8310	10863	339	54293
		(100.00)	(1.99)	(24.50)		(8.36)	(1.87)	(10.31)	(47.02)	(28.22)	(1.71)	(7.15)	(9.35)	(0.29)	(46.73)
	1990-91	123593	2510	28581	N.G.	11631	4658	14193	61523	26641	6621	8510	10894	694	57360
		(100.00)	(202)	(23.08)		(9.41)	(3.77)	(11.48)	(49.78)	(21.56)	(5.36)	(6.89)	(12.05)	(0.56)	(46.41)
	1995-96	123646	1490	24900	449	9613	5100	12623	54193	41250	3755	8009	12189	285	65488
		(100.00)	(1.21)	(20.14)	(0.36)	(7.79)	(4.12)	(10.21)	(43.83)	(33.36)	(3.04)	(6.48)	(9.36)	(0.23)	(52.96)
Kallam															
	1982-83	97769	1717	28369	290	9613	3775	9521	53285	12580	7604	5358	6360	350	32252
		(100.00)	(1.76)	(29.02)	(0.30)	(9.83)	(3.86)	(9.74)	(54.50)	(12.87)	(7.78)	(5.58)	(6.51)	(0.36)	(32.99)
	1986-87	97456	1030	25925	899	270	1250	2655	32029	10000	3600	6300	8677	N.G.	28577
		(100.00)	(1.06)	(26.60)	(0.92)	(0.28)	(1.28)	(2.72)	(32.87)	(10.26)	(3.69)	(6.46)	(8.90)		(29.32)
	1990-91	112516	1772	26267	13543	11300	8000	12836	73718	13543	4143	9893	15515	161	43255
		(100.00)	(1.57)	(23.44)	(12.04)	(10.04)	(7.11)	(11.41)	(65.52)	(12.04)	(3.68)	(8.79)	(13.79)	(0.14)	(38.44)
	1995-96	96078	1482	20692	3682	9361	2060	17587	54864	12776	1255	3293	11459	N.G.	28783
		(100.00)	(1.54)	(21.54)	(3.83)	(9.74)	(2.14)	(18.30)	(57.10)	(13.30)	(1.31)	(3.43)	(11.93)		(29.96)
Omerga															
	1982-83	133005	2124	28091	2103	13929	1678	17942	65867	28210	4595	6700	3822	1226	44553
		(100.00)	(1.60)	(21.12)	(1.58)	(10.47)	(1.26)	(13.49)	(49.52)	(21.21)	(3.45)	(5.04)	(2.87)	(0.92)	(33.50)
	1986-87	150299	21.50	35268	2110	12540	1756	24208	78032	21410	4620	5950	5068	1206	45254
		(100.00)	(1.43)	(23.47)	(1.40)	(8.34)	(1.17)	(16.11)	(51.92)	(18.90)	(3.07)	(3.96)	(3.37)	(0.80)	(30.11)
	1990-91	154067	4010	24150	1000	22420	2275	27477	81332	21405	3639	9950	17308	1033	53335
		(100.00)	(2.60)	(15.67)	(0.65)	(14.55)	(1.48)	(17.85)	(52.79)	(13.89)	(2.36)	(6.46)	(11.23)	(0.67)	(34.62)

Contd....

Talukas	Year	Gross Cropped Area	Kharif Crops							Rabi Crops					
			Paddy	Jowar	Bajra	Tur	Sugar-cane	Oil-Seeds	Total Kharif	Jowar	Wheat	Gram	Oil-Seeds	Fruits & veg.	Total Rabi
	1995-96	150342	5620	19905	3040	18460	3075	32139	82239	28310	4820	8490	13960	459	56039
		(100.00)	(3.74)	(13.24)	(2.02)	(12.28)	(2.05)	(21.38)	(54.70)	(18.83)	(3.21)	(5.65)	(9.29)	(0.31)	(37.27)
Tuljapur															
	1982-83	84007	6063	11569	2316	11240	5984	6513	43685	12664	4033	5974	2445	712	25828
		(100.00)	(7.22)	(13.77)	(2.76)	(13.38)	(7.12)	(7.75)	(52.00)	(15.07)	(4.80)	(7.11)	(2.91)	(0.85)	(30.75)
	1986-87	113699	4135	6200	1051	12391	2112	16150	42039	27075	1472	4509	18344	7099	58499
		(100.00)	(3.64)	(5.45)	(0.92)	(10.90)	(1.86)	(14.20)	(36.97)	(23.81)	(1.29)	(3.97)	(16.13)	(6.24)	(49.45)
	1990-91	147998	5944	16694	3158	20450	6983	17454	70683	27298	4423	4722	18751	6653	61847
		(100.00)	(4.02)	(11.28)	(2.13)	(13.82)	(4.72)	(11.79)	(47.76)	(18.44)	(2.99)	(3.19)	(12.67)	(4.50)	(41.79)
	1995-96	147540	6836	17819	136	18980	4175	27835	75781	27810	9413	4616	18730	7344	67913
		(100.00)	(4.63)	(12.08)	(0.09)	(12.86)	(2.83)	(18.87)	(51.36)	(18.85)	(6.38)	(3.13)	(12.69)	(4.98)	(46.03)
Paranda															
	1982-83	94538	540	450	260	7220	570	2390	11430	54670	4510	3320	12240	185	74925
		(100.00)	(0.57)	(0.48)	(0.28)	(7.64)	(0.60)	(2.53)	(12.9)	(57.83)	(4.77)	(3.51)	(12.95)	(0.20)	(89.25)
	1986-87	109484	588	N.G.	1443	8173	1540	12736	24470	54808	3175	7565	18094	330	83972
		(100.00)	(0.54)		(1.31)	(7.45)	(1.41)	(11.63)	(22.35)	(50.06)	(2.90)	(6.91)	(16.83)	(0.30)	(76.70)
	1990-91	125369	467	708	1839	8164	2200	13056	26434	76729	3003	3974	12345	688	96739
		(100.00)	(0.37)	(0.56)	(1.47)	(6.51)	(1.75)	(10.41)	(21.08)	(61.20)	(2.40)	(3.17)	(9.85)	(0.55)	(77.16)
	1995-96	124800	N.G.	N.G.	298	10241	1184	11610	23333	77959	5011	8139	9562	329	101000
		(100.00)			(0.24)	(8.21)	(0.95)	(9.30)	(18.70)	(62.47)	(4.02)	(6.52)	(7.66)	(0.26)	(80.93)
Bhoom															
	1982-83	66919	1068	3955	290	4022	198	3530	13063	33204	4327	5023	9203	474	52231
		(100.00)	(1.60)	(5.91)	(0.43)	(6.01)	(0.30)	(5.28)	(19.52)	(49.62)	(6.47)	(7.51)	(13.75)	(0.71)	(78.05)
	1986-87	67858	472	11736	1795	6787	479	8171	29440	16867	2997	3106	7767	331	31068
		(100.00)	(0.70)	(17.29)	(2.65)	(10.00)	(0.71)	(12.04)	(43.38)	(24.87)	(4.42)	(4.58)	(11.45)	(0.49)	(45.78)
	1990-91	76332	833	10178	1300	9700	800	1556	38367	14493	3918	3371	10106	277	32165
		(100.00)	(1.09)	(13.33)	(1.70)	(12.71)	(1.05)	(20.38)	(50.26)	(18.99)	(5.13)	(4.42)	(13.24)	(0.36)	(42.14)
	1995-96	95437	766	11333	3607	11617	1357	13096	75781	25366	5424	4169	8567	369	43885
		(100.00)	(0.80)	(11.87)	(3.78)	(12.17)	(1.42)	(13.72)	(51.36)	(26.58)	(5.68)	(4.37)	(6.98)	(0.38)	(45.98)

Contd....

Talukas Year	Gross Cropped Area	Kharif Crops							Rabi Crops					
		Paddy	Jowar	Bajra	Tur	Sugar-cane	Oil-Seeds	Total Kharif	Jowar	Wheat	Gram	Oil-Seeds	Fruits & veg.	Total Rabi
District Total														
1982-83	584076	13912	95731	5589	55841	14237	44517	229887	171543	32854	34965	43323	3252	286937
	(100.00)	(2.38	(16.39)	(0.96)	(9.56)	(2.254)	(7.73)	(39.36)	(29.37)	(5.62)	(5.99)	(7.42)	(0.56)	(48.95)
1986-87	654981	10685	107590	7288	49871	9312	75895	260641	169951	17854	35740	68813	9305	311663
	(100.00)	(1.63)	(16.43)	(1.11)	(7.61)	(1.42)	(11.59)	(39.79)	(25.94)	(2.73)	(5.45)	(10.51)	(1.42)	(46.06)
1990-91	739871	15536	106528	20840	83665	24916	100572	352057	180109	25747	40420	88919	9596	344701
	(100.00)	(2.10)	(14.40)	(2.82)	(11.31)	(3.37)	(13.59)	(47.58)	(24.34)	(3.48)	(5.46)	(12.03)	(1.28)	(46.59)
1995-96	737843	16194	94649	11212	78290	16951	114890	332186	213471	29678	36716	74467	8776	363108
	(100.00)	(2.19)	(12.83)	(1.52)	(10.61)	(2.30)	(15.57)	(45.02)	(28.93)	(4.02)	(4.98)	(10.09)	(1.19)	(49.21)

N.B.: Figures in the brackets indicate the percentage to the GrossCropped Area.
N.G. Shows not grown
Source: Estimated on the basis of Data collected from Socio-Economic Review and District Statistical Abstract of Osmanabad District, Directorate of Economics and Statistics, Government of Maharashtra, 1985-86, 1989-90,1990-91 and 1998-99.

Table 7
Percentage change in Area Under Important Crops in Osmanabad District (1982-83 to 1995-96)

Talukas	Kharif Crops							Rabi Crops					
	Paddy	Jowar	Bajra	Tur	Sugar-cane	Oil-Seeds	Total Kharif	Jowar	Wheat	Gram	Oil-Seeds	Fruits & veg.	Total Rabi
Osmanabad	-(1.02)	-(1.46)	+(0.05)	-(1.30)	+(2.24)	+(5.87)	+(4.37)	+(5.35)	-(4.18)	-(1.49)	+(1.28)	-(0.05)	+(0.91)
Kallam	-(0.22)	-(4.47)	+(3.53)	-(0.09)	-(1.71)	+(8.56)	+(2.60)	+(0.42)	-(6.47)	-(2.01)	+(5.31)	-(0.36)	-(3.03)
Omerga	+(2.14)	-(7.88)	+(0.44)	+(1.81)	+(0.69)	+(7.79)	+(5.18)	-(2.38)	-(0.24)	+(0.61)	+(6.41)	-(0.62)	+(3.77)
Tuljapur	-(2.59)	-(1.69)	-(2.67)	-(0.52)	-(4.29)	+(11.12)	-(0.64)	+(3.78)	+(1.57)	-(3.98)	+(9.78)	+(4.13)	+(15.28)
Pranda	-(0.57)	-(0.20)	-(0.03)	+(0.57)	+(0.34)	+(6.77)	+(6.60)	+(4.64)	-(0.75)	+(3.01)	-(5.29)	+(0.07)	+(1.67)
Bhoom	-(0.79)	+(5.96)	+(3.35)	+(6.15)	-(1.14)	+(8.45)	+(24.25)	-(23.04)	-(0.79)	-(3.14)	-(4.77)	-(0.33)	-(32.60)
District	-(0.19)	-(3.56)	+(0.56)	+(1.05)	-(0.14)	+(7.94)	+(5.66)	-(0.44)	-(1.60)	-(1.01)	+(2.67)	+(0.73)	+(0.26)

N.B: Figures in the brackets Shows change inpercentange of Area Under Different Crops to the Gross Cropped Area.
Source: Estimated on the basis of Data collected form Socio - Economic Review and District Statistical Abstract of Osmanabad District, Directorate of Economics and Statistics, Government of Maharashtra, 1985-86, 1989-90, 1990-91 and 1998-99.

occupied (15.57 per cent) highest cropped area under kharif season as it was the most favoured crop in the district followed by kharif jowar (12.83 per cent, tur (1061 per cent), sugarcane (2.30 per cent), paddy (2.19 per cent) and bajra (1.52 per cent). There was marginal increase in the cultivated area under oilseeds (7.94 per cent), tur (9.56 per cent) and bajra (0.56 per cent), and decrease in the area under kharif jowar (-356), paddy (-0.19 per cent) and sugarcane (-0.14 per cent) during last fourteen years in the study district. From this, it is seen that there was very high increase in percentage of area under oilseeds (7.94 per cent), and a very high decrease in percentage of area under kharif jowar i.e (-3.56 per cent) and small decrease in percentage area under paddy (-0.19 per cent) and sugarcane (-0.14 per cent).

A perusal of tables 6 and 7 shows that, during the last fourteen years, the percentage share of kharif cultivated area to gross cropped area in the district has increased by 5.66 per cent. There was a gradual increase in the area under kharif crops in the district as a whole during the last fourteen years. This increase was seen in the five talukas, namely, Bhoom (24.25 per cent), Paranda (6.60 per cent), Omerga (5.18 per cent), Osmanabad (4.37 per cent) and Kallam (2.60 per cent) and decrease (-0.64 per cent) in Tuljapur taluka.

PADDY

Paddy is an aquatic plane which requires high rainfall or assured irrigation facilities. Paddy occupied 2.19 per cent (1995-96) of the gross cultivated area in Osmanabad district. Cultivated area under paddy has increased by 16 per cent during the last fourteen years. But its percentage share in gross cropped area has registered a decrease of -0.19 per cent. The area under paddy has registered an increase in Omerga taluka by 2.14 per cent, significant increase in paddy cultivation is found in Omerga taluka where the irrigation facilities were developed. The area under paddy has registered a decrease in Tuljapur (-2.59 per cent), Osmanabad (-1.02 per cent), Bhoom (-0.79 per cent), Kallam (-0.22 per cent) and paranda (-0.20 per cent) talukas.

JOWAR (KHARIF)

Jowar (Kharif) can withstand drought to a considerable extent and is grown both as dry as well as an irrigated crop. It is a staple food of this study area and is also used as fodder. But Kharif jowar is denoted as an inferior food compared to rabi jowar. Jowar (Kharif) was the important crop during 1982-83 as it occupied 16.39 per cent of the gross cultivated area.

A perusal of table 6 shows that, in 1982-83, jowar occupied (16.39 per cent) highest percentage share of gross cropped area in the district. During last fourteen years, there was decrease (-3.56 per cent) in area under kharif jowar. During the last fourteen years, the talukas of Omerga (-7.88 per cent),Kallam (-7.47 per cent) ,Osmanabad (-1.46 per cent),Tuljapur (-1.69 per cent), and paranda (-0.57 per cent) have recorded decrease in jowar cultivated area. Only Bhoom taluka has recorded the highest (5.96 per cent) increase in area under jowar in the district.

BAJRA

Bajra was a drought resistant crop requiring less amount of rainfall. From the table 6, it is observed that during 1982-83, there was only 0.96 per cent of the gross cropped area under bajra in the district. Bajra crop occupied 1.52 per cent of the gross cropped area of the study region in 1995-96. It has registered an increase of 0.58 per cent over a period of fourteen years. The area under bajra has registered increase in the talukas of Bhoom (3.35 per cent), Kallam (3.53 per cent) ,Omerga (0.44 per cent) and Osmanabad (0.05per cent) and it has registered decrease in two talukas of Tuljapur (-2.67 per cent) and paranda (-0.03 per cent).

TUR

Tur was the important pulse crop and it occupied 9.56 per cent of gross cultivated area of the district in 1982-83. During 1995-96, it occupied 10.61 per cent of gross cultivated area of the Osmanabad district. Thus, it has registered an increase of 1.05 percentage points in cultivated area Area under tur has registered an increase in three talukas, namely, Bhoom (6.15 per cent) omerga (1.81 per cent) and paranda (0.57 per cent). While it has registered decrease in area in Osmanabad (-1.30 per cent), Tuljapur (-0.50 per cent) and Kallam (-0.09 per cent) talukas.].

SUGARCANE

Sugarcane was an important cash crop and it occupied 2.44 per cent (1982-83) of the gross cultivated area of the district under study. Similarly, sugarcane occupied 2.30 per cent of the gross cropped area of the district in 1995-96. But it registered decline in cultivated area (-0.14 per cent) during the last fourteen years. This crop registered a decrease in area in two talukas, namely, Tuljapur (-4.29 per cent) and Kallam (-1.71 per cent). But this crop has registered an increase in area in remaining talukas, namely, Osmanabad (2.24 per cent), Bhoom (1.14 per cent), Omerga (0.79 per cent) and Paranda (0.37 per cent) during the period under study.

OILSEEDS

Groundnut, sunflower and Sesamum are the important oilseed crops grown in kharif season which are mostly dependent on irrigation facilities at the time of rainless period in the study area. It is seen from the table

6 and 7 that, kharif oilseeds occupied 7.63 per cent (1982-83) of the gross cultivated area of the district. And it was 15.57 per cent of the gross cultivated area in 1995-96. Thus, oilseeds have registered an increase in cultivated area of 7.94 per cent points during the last fourteen years. It has registered an increase in area in almost all talukas of the district. Tuljapur taluka has registered the highest increase (11.12 per cent) in area under kharif oilseeds, followed by Kallam (8.56 per cent), Bhoom (8.45 per cent), Omerga (7.89 per cent), Paranda (6.77 per cent) and Osmanabad (5.87 per cent) talukas. There was increase in cultivated area under oilseed crops during last fourteen years in the study region.

Crop Pattern Effect of Irrigation: Rabi Season

The rabi season starts after the harvest of kharif crops in September-October. The rabi crops are fine grained and require a good amount of irrigation water. Since this is not a rainy season and winter rains are not common, water is obtained mainly from man-made sources of irrigation. The important rabi crops grown in this district viz. jowar, wheat, gram, oilseeds, fruits and vegetables, are considered in this study.

A perusal of table 6 shows that in 1982-83 cropwise cultivated area of rabi season of the whole district was 2,85,937 hectares and in 1995-96 it was 3,63,108 hectares. Thus during last fourteen years (from 1982-83 to 1995-96), there was 27 per cent increase in rabi cultivated area. Similarly a perusal of table 6 and 7 shows that, the percentage share of rabi cultivated area to gross cropped area in the district was 48.95 per cent in 1882-83 and 49.51 per cent in 1995-96. The percentage share of rabi cropped area to gross cropped area in the district has increased by 1.26 per cent over the period of last fourteen years.

Thus, it is seen from the tables 6 and 7 that, during the last fourteen years percentage share of rabi cultivated area to gross cultivated area in Osmanabad district has increased by 1.26 per cent. In most of the talukas, the percentage share of area under rabi crops has registered an increase of 15.28 per cent in Tuljapur, 3.77 per cent in Omerga, 1.67 per cent in Paranda and 0.91 per cent in Osmanabad talukas. Rest of the talukas, namely, Bhoom (-32.07 per cent) and Kallam (-3.03 per cent) have recorded a decrease in rabi cultivated area.

RABI CROPS

Jowar

Jowar was the most important crop in 1982-83, which occupied 29.37 per cent of the total cropped area in the study district. In 1995-96, again it occupied 28.93 per cent of the total cropped area of the district. It has registered a small decrease of -0.44 per cent. But most of the talukas have recorded increase in area under rabi jowar over the period under study. Osmanabad taluka has recorded the highest (5.35 per cent) increase in area under jowar followed by Paranda (4.64 per cent), Tuljapur (3.78 per cent) Kallam (0.42 per cent) and Omerga (-2.38 per cent) have recorded decrease in area under rabi jowar. Thus, during last fourteen years, area under rabi jowar crop in the district has increased.

Wheat

In 1982-83, wheat occupied 5.62 per cent of gross cultivated area of the district, while in 1995-96, wheat covered an area of 4.02 per cent of the gross cropped area of the district. During the last fourteen years, the cultivated area under wheat decreased by 1.60 percentage points. Only Tuljapur taluka has recorded an increase of 1.57 per cent in cropped area and 6.09 per cent in irrigated area and thus recorded a significant increase in wheat crop. Remaining five talukas have recorded a decrease in cropped area and irrigated area under wheat over the period under study. This crop has lost its importance and there was shift in its area to other cash crops like oilseeds.

Gram

A perusal of tables 6 and 7 shows that, gram covers an area of 5.99 per cent of the gross cropped area of Osmanabad district in 1982-83. It occupies 4.98 per cent of the district's gross cropped area in 1995-96. Thus after a lapse of fourteen years, percentage share of area under gram to gross cropped area in the district has decreased by 1.01 per cent. Tuljapur taluka (-3.98) has recorded the highest decrease in area under gram followed by Bhoom (-3.14 per cent), Kallam (-2.01 per cent) and Osmanabad (-1.49 per cent) talukas. But the talukas such as the Paranda (3.01 per cent) and Omerga (0.61 per cent) have recorded an increase in area under gram.

Oilseeds (Rabi)

Among the total rabi oilseeds grown in this study area, safflower, linseed, sunflower and groundnut are the important oilseed crops which covered 7.42 per cent (1982-83) of the gross cropped area of the district. While in 1995-96, these crops occupied 10.09 per cent of the gross cropped area of the district. Thus, during last fourteen years, there was 2.67 percentage points increase in area under rabi oilseeds. So during last fourteen years, it is observed that, there was an increase of 2.67 per cent in its area. Similar increase was also observed in Tuljapur (9.78 per cent), Omerga (6.41 per cent), Kallam (5.31 per cent) and Osmanabad (1.28 per cent) talukas. Rest of the talukas namely, Paranda (-5.29 per cent) and Bhoom (-4.77 per cent) have recorded a decline in area under rabi oilseeds. Thus, it is observed that farmers are increasingly shifting from the cultivation of wheat and gram to oilseeds cultivation as they find it more remunerative and high yielding crop. Oilseed crops require less amount of watering compared to other crops in rabi season. From the point of view of the farmers, the growing of oilseeds is more remunerative than the growing of food grains.

During last fourteen years it is observed that, with the development of irrigation facilities, the farmers of this district gave preference to rabi oilseeds as compared to food grain crops. Thus, increasing irrigation has positive impact on oilseed crops.

Fruits and Vegetables

Among the fruits and vegetables grown in Osmanabad district, banana, grapes, mango and onion are the important crops. Fruits like banana and grapes require both more watering and money expenditure. A perusal of tables 6 and 7 shows that fruits and Vegetables have occupied 0.56 per cent of the gross cropped area of the district in 1982-83, while in 1995-96, these crops covered an area of about 1.19 per cent of gross cropped area. Thus, during last fourteen years, the cultivated area under this crop has increased by 1.19 per cent. Similarly, percentage of area under fruits and Vegetables to gross cropped area under these crops has also increased by 1.76 per cent over the period in the district. Two talukas, namely, Tuljapur and Paranda have recorded highest increase in cultivated area. The cultivated area under fruits and Vegetables in this district has increased and this increase is attributed to the increase in irrigation facilities. Therefore, development of irrigation in the study area has positive impact on fruits and Vegetables. Thus, irrigation development was in favor of farmers of Osmanabad district in diversification of crops from the traditional cereal crops to commercial crops. Therefore, growth of irrigation in the district has a positive impact on commercial crops.

IMPACT OF IRRIGATION ON YIELD

Irrigation water alone cannot push the yield of crops. The quantity of produce is a function of several variables and irrigation is only one of them. But irrigation is catalytic factor and on the availability of which depends the use of other inputs like fertilizers, high yielding varieties of seeds etc. This change in the character of cultivation under irrigation leads to greater productivity of crops previously cultivated. Such improvements in the per hectare yield are technology induced and irrigation constitutes the major causative factor behind such yield rate enhancement. Studies on the relative contribution of various component elements in the growth of agricultural output by Kalhon and Johl, Minhas and Vaidyanathan Prabha Chander have established that increase in per hectare yield is attributable to the adoption of new agricultural technology emerging from rapid expansion of irrigation facilities.[10]

Dhawan has studied various issues relating to the impact of irrigation on productivity with reference to Indian farm economy. His study reveals that the total output expansion in crop output per irrigated hectare was nearly 19 quintals in food grain energy equivalents (FEEs units) during mid-eighties. Under rainfed condition, country's crop output per hectare was 8 quintals in FEEs units.[11] Sharma has attempted to study the socio-economic impact of major irrigation projects in India. The study has found that average yields of all crops are substantially higher (8 to 500 per cent) in "after" project situation compared to before project situation.[12]

We may now examine irrigation impact on yields of important crops in Osmanabad district. The year wise yields of important crops and the percentage change in yields of important crops in the study area during 1982-83 to 1996-97 have been calculated and are presented in table 8.

Table 8
Cropwise Yields and Precentage Change in Yields of Main Crops in Osmanabad District (1982-83 to 1996-97)

(Crop yields in kg/ha, sugarcane yields in dressed cane/ha and fruit yields in tonnes /ha)

Crops	*Year - wise yields*				*Percentage change in yields during 1982-83 to 1996-97*
	1982-83	*1986-87*	*1990-91*	*1996-97*	
Kharif Crops					
1. Paddy	627	131	959	700	11.64
2. Jowar	1074	209	1022	1431	33.24
3. Bajra	379	192	404	762	101.05
4. Tur	390	117	272	601	54.01
5. Sugarcane	82610	56985	74020	61877	-25.10
6. Groundnut	690	112	567	478	-30.72
7. Sunflower	647	170	576	665	2.78
Rabi Crops					
1. Rabi Jowar	433	261	770	893	106.24
2. Wheat	721	387	959	1000	38.70
3. Gram	404	149	509	586	45.05
4. Safflower	655	433	328	587	-10.38
5. Groundnut	1234	1734	800	1476	19.61
6. Sunflower	488	510	581	652	33.61
7. Banana	16.64	–	–	53.81	223.38
8. Grapes	11.40	–	12.60	35.80	214.04
9. Onion	9.42	–	10.00	12.00	27.40

Source: Districtwise Agricultural Statistical Infromation of Maharashtra, Part II, Commissionerate of Agricultural, Pune, 1986-87, 1995-96 and 1997-98.

Kharif Crops

It is seen from the table 8 that, during last fifteen years (1982-83 to 1996-97), there was considerable increase in yields of kharif crops in Osmanabad district.

Among the kharif crops, bajra crop recorded highest increase in yields (101.05 per cent) in the district followed by tur (51.10 per cent), kharif jowar (33.24 per cent), paddy (11.64 per cent), sunflower (2.78 per cent) and there was decrease in yields of sugarcane (-25.10 per cent) and groundnut (-30.72 per cent) during last fifteen years. It shows that any coarse grains have made drastic improvement in their yields. This is mainly due to the fact that, high yielding variety of coarse grains has come a bit later, they are drought resistant. Being a kharif or monsoon crop, they depend considerably on rainwater and they do not require much irrigation water. While sugarcane requires plenty of irrigation water and it cannot resist even a slight drought. Consequently, there is hardly any unirrigated sugarcane. Sugarcane cultivation with the help of perennial source of irrigation is bound to give high yields. Similarly, paddy requires a lot of irrigation water but being a kharif crop, there may not be much irrigation water from artificial source of water. Rest of the crops like kharif jowar, bajra, sunflower, groundnut depend mainly on rainwater but during the drought period, these crops require irrigation water, which is useful for increase in crop yields.

Rabi Crops

A perusal of table 8 shows that during the last fifteen years (1982-83 to 1996-97) the percentage increase in yields of rabi crops have been comparatively higher than that of kharif crops. Banana recorded highest increase of 223.38 percentage in yields during last fifteen years, followed by grapes (214.04 per cent), rabi jowar (106.24 per cent), gram (45.05 per cent), wheat (38.70 per cent), sunflower (33.61 per cent), onion (27.40 per cent) groundnut (19.61 per cent), and only safflower recorded decrease or negative change (-10.38 per cent) in its yield during last fifteen years in Osmanabad district. This may be due to the availability if irrigation facilities and high yielding varieties of jowar and the farmers preference of giving full inputs to this grain because this is most favorable crop and farmers get more returns from it. There was considerable increase in yields of wheat and gram also. There was also increase in the yields of oilseeds in Osmanabad district. Fruits like banana and grapes recorded highest increase in yields in Osmanabad district. This may be due to the farmers preference for giving full inputs to the fruits as this is highly remunerative crop compared to other crops. All rabi crops recorded positive changes during last fifteen years in Osmanabad district. Most of the rabi crops depend on irrigation water as compared to kharif crops.

Thus, it can be concluded that, during last fifteen years (1982-83 tp 1996-97), irrigation has helped for increasing yields of kharif and rabi crops in Osmanabad district.

CROP PRODUCTION EFFECT OF IRRIGATION

It is now understandably an established fact that irrigation is expected to modernize agriculture where it not only brings larger area under cultivation but also stimulates and increases agricultural production also. The interaction effect of irrigation in Osmanabad district, i.e. the expansion of the aggregate volume of agricultural crop output resulting from extension of net area under cultivation increase in intensity of farming suitable shifts in crop pattern and higher productivity in crops per unit of land may now be examined. It is well known fact that with the development of irrigation facilities the agricultural production has also increased. Table 9, which seeks to highlight this phenomenon presents increase in total production of all crops with increase in irrigated area in the district during the period 1984-85 tp 1996-97. So it was worth

Table 9
Correlation Between Total Irrigated Area and Total Crop Production, Irrigation Area Under Food-grains and Food-grains Production and Irrigation Area Under Non-food-grains and Non-food-grains Production in Osmanabad District (1984-85 to 1995-96)

(Area in hectares and production in metric tonnes)

Year	*Total irrigated area*	*Total crop production*	*Irrigated Area Under Food-grains*	*Food-grains production*	*Irrigated Area Under non food grains*	*Non food-grains production*
1984-85	94218	4682	65508	3247	28710	1435
1985-86	95835	3036	69211	1991	26624	1045
1986-87	101978	5325	66457	968	35521	4357
1987-88	121327	9128	74757	3289	46570	5839
1988-89	146576	10781	80323	2333	66253	8448
1989-90	138282	14140	75393	3497	62889	10643
1990-91	152403	19529	82428	3577	69775	15952
1991-92	135397	13033	70498	1526	74900	11507
1992-93	115924	10636	67240	3505	486874	7131
1993-94	112522	11672	67953	3952	44569	7720
1994-95	121278	17326	68932	2724	52346	14602
1995-96	122071	14258	70182	3321	51889	10937
Mean standard	1041293	9541.00	61348.64	2423.57	42780.71	7115.43
Deviation	18961.13	5016.83	5473.97	927.98	14499.54	4710.61
R	0.79**		0.19		0.85***	

**Significant at 5 percent level
***Significant at 1 percent level

Source: Estimated from (1) The data collected from Socio-Economic Review and District Statistical Abstract of Osmanabad District, Directorate of Economics and Statistics, Government of Maharashtra, 1985-86, and 1998-99. (2) District-wise Agricultural Statistical Information of Maharashtra Part II, Commissionerate of Agricultural, Pune, 1986-87, 1995-96 and 1997-98.

while to see the correlation between total irrigated area and total crop production, irrigated area under food-grains and food-grain production and irrigated area under non food-grains and non food-grains production. Taking year-wise data from 1984-85 to 1995-96 for Osmanabad district, correlation coefficients were computed and are presented in table 9.

The table 9 reveals that total crop production, food grains production and non-food grain production in the district have increased with the increase in irrigated area since 1984-85. While during 1984-85 to 1995-96 the mean of the total production, food grain production, non food grain production in the district were 9539, 2423.57 and 7115.43 tonnes respectively. A perusal of table 9 shows that the correlation co-efficient between total irrigated area and total crop production during 1984-85 to 1995-96 in the district was 0.79 and this positive correlation co-efficient is found to be statistically significant at 5 per cent level. It has been observed that the district with more irrigation facilities during a period of twelve years recorded high correlation mainly due to irrigation facilities which led to the use of high yielding varieties of seeds and fertilizers and therefore, resulted in high crop production. The result shows a high positive correlation coefficient between total irrigated area and total crop production. This suggests that irrigation is very important ingredient for increasing agricultural production It is the fact that as irrigation increases the use of other inputs like fertilizers, high yielding varieties of seeds also increases.

The correlation co-efficient between irrigated area under food grains and food grain production (including cereals and pulses) in the district was 0.19 and it is found to be statistically insignificant. This shows that during the last twelve years, irrigation development was in favour of farmers of the district in diversification of crops from the traditional cereal crops to commercial crops. It is seen from the table 8 that most of the land of Osmanabad district during the kharif and rabi seasons was under jowar cultivation and after lapse of fourteen years (1982-83 to 1995-96), Jowar (rabi) remained the first ranking crop. However, there was decrease in rabi jowar by -0.44 per cent and shift in first ranking crop from jowar (kharif) to oilseeds in kharif season. This shows that there has been diversification of crops from the traditional cereal crops to commercial crops in the district.

The correlation coefficient between irrigated area under non food grains and non food grain production (including oilseeds, sugarcane, fruits and vegetables) in the district recorded positive correlation (0.85) and this coefficient is found to be statistically significant at 1 per cent level. After a lapse of twelve years (1984-85 to 1995-96) (see tables 8 and 9) there was shift in first ranking crop from jowar (kharif) to oilseeds in Kharif season and oilseeds were remained as the second ranking crop in rabi season. Therefore the results show a high positive correlation between irrigated area under non-food grains and production of non-food grains in the district. This suggests that irrigation has the dominant role in non-foodgrain production. It is the fact that as irrigation increases the use of other inputs like fertilizers, high yielding varieties of seeds also increase. It may, therefore, be concluded that during the last twelve years (1984-85 to 1995-96)because of expansion and intensive utilization of irrigation facilities, the Osmanabad district has made rapid strides in agricultural production. This shows that irrigation development in the Osmanabad district has beneficial production effect.

Thus from the macro-level analysis, it can be concluded that extension of irrigation facilities has made positive effect on agriculture in Osmanabad district. The growth of irrigation in the district has positive effect on cultivated area. The irrigation development in Osmanabad district has beneficial area effect, intensity effect, crop pattern effect, yields effect and production effect. Thus the impact of irrigation on agriculture in Osmanabad district is favourable.

The study was intended to objectively evaluate the major macro level impacts of irrigation on agriculture in the study area.

On the whole, the benefits of irrigation development have been found to be positive in the study area. the analysis of area effects of irrigation showed that during last fourteen years (i.e. 1982-83 to 1995-96)there was gradual increase (4.18 per cent)in net cultivated area 20.85 per cent increase in both double and gross cropped area in Osmanabad district. The area effects of irrigation have been significant with increase in the index of crop intensity resulting from double and multiple cropping. The crop intensity index has increased from 104.08 per cent to 124.93 per cent during 1982-83 to 1995-96. The "Crop pattern effect" of irrigation has been favourable in the study area since the irrigation has helped the farmers in shifting low value crops to high value crops. During last fourteen years (1982-83 to 1995-96) there was considerable increase in per hectare crop yields of kharif and rabi crops in the study area. The beneficial role of irrigation in bringing stability to crop output is observed. A positive correlation between total irrigated area under non-foodgrains and non-foodgrain production is observed in the district. But correlation between irrigated area food grain and food-grains production is found to be in significant.

Thus, different impacts of irrigation on agriculture have been found to be positive in the district under study.

NOTES AND REFERENCES

1. Bhagirath, vol xxxx vi, April-June 1999, Central water commission and ministry of water resources, New Delhi, pp 78-82.
2. Ghosh K.N., Agriculture in Economic Development with special reference to Punjab, Vikas Publications, New Delhi, 1977, p. 63.

3. Bhardwaj Krishna; Production Conditions in Indian Agriculture, A Study Based of Farm Management Survey, Cambridge University Press, 1974, p. 43.
4. Gadgil D.R. Economic Effects of Irrigation, Report of Survey of the Direct and Indirect Benefits of Godavari and Pravara Canals, Pub. No. 17, Gokhale Institute of Economics and Polities, Pune, 1948, p. 22.
5. Mukhopadhyaya S.K., Soures of Variation in Agricultural Productivity, Macmillan Company of India, 1974, p. 64.
6. Singh J.P. et al., New Seeds, Adoption and Yields, Institute of Social and Economics Change, Ahmedabad, 1979, p. 17. and Gaikwad V.R. and Desai V.M. Development of Incentive Agricultural Lessons From Intensive Agricultural District Program, Indian Institute of Management, Ahmedabad, p. 12.
7. Ghosh R.N., op. cit., p. 63.
8. Mishra S., A Comparative Study of the Economics of Minor Sources of Irrigation in Uttar Pradesh (A Report), Oxford and I.B.H. Publishing Co., 1968, p. 39.
9. Bhardwaj Krishna, op. cit., p. 48.
10. Kalhon A.S. and Johl S.S. Agricultural Growth Rates in Punjab, Punjab Agricultural University, 1968.

    Minhas B.S. and Vaidyanathan A., Growth of Crop Output in India, 1951-54 to 1958-61., An Analysis by Component Elements, *Journal of the Indian Society of Agricultural Statistics*, Vol. XVII, 1965.

    Prabha Chander, Districtwise Measurement and Decomposition of Agricultural Output in Panjab, *Indian Economic Review*, Vol. VI, 1971.
11. Dhawan B.D., Irrigation in India's Agricultural Development-Productivity, Stability and Equity, Sage Publishers, New Delhi, 1988, p. 26.
12. Sharma S.D., Socio-Economic Impacts of Major Irrigation Projects, Proceedings of Fifty Eighth Annual Research and Development Session, Vol II, Central Board of Irrigation and power, Bangalore, 1993.

D.P. TAKALE

RESOURCE PRODUCTIVITY, RETURNS TO SCALE AND RESOURCE-USE EFFICIENCY ON IRRIGATED FARMS

7

INTRODUCTION

India is a developing country. Agriculture is the main occupation of the people of India as about 63 per cent of the population depends directly on it. Indian agriculture contributes to economic growth. The growth of agricultural sector depends on various factors such as manures and fertilizers, irrigation facilities, manpower, capital, improved seed, new technology etc. The main problem of increasing agricultural production in any area is as to how best to increase the output per unit of input. Farmers have limited resources and their objective is to maximize farm income form the available resources. In Indian agriculture, the availability of resources is not same all over the country. Some of the crucial resources are either under-utilized or over-utilized in the crop production. Therefore, the optimum allocation of resources is necessary for the growth of agricultural production.

For increasing farm income, the adjustment and efficient use of farm resources is important. If the farmers are efficient in the use of scare resources then farmers can increase their production by making adjustments in the use of factors of production in optimal direction. In case, it is efficient the only way out for increasing production would be the adoption of modern inputs and improved technology of production. Increasing cost of agricultural resources by quantum jumps has been making agriculturel a losing proposition. In view of this, it is necessary that the available resources should be used economically and efficiently. Therefore, the present study is to focus on the resource-use efficiency on irrigated farms in the command area of Purna Irrigation Project of Maharashtra state.

CONCEPTS OF RESOURCE-USE EFFICIENCY

Economics is in many ways a science of efficiency. But the concept of economic efficiency is significant to determine the cost per unit of output.

Farrel has divided the economic efficiency into two components, namely, (1) technical efficiency, and (2) the price efficiency .

Technical Efficiency

This type of efficiency refers to the proper choice of production function among all those actively in use by farms in the agriculture. This efficiency depends on marginal ability, soil fertility, climatic conditions, price policy, market condition, incentives etc.

Price Efficiency

The price efficiency refers to the proper choice of input combination. Price efficiency is also known allocative efficiency. Allocative efficiency refers to the achievement of optimum output so as to maximize net income.

Economic efficiency combines both technical and allocative efficiency. A technically efficient farm need not be a price efficient farm.

The problem of empirical measurement of efficiency is difficult when the entire agricultural industry is to be viewed from the standpoint of economic efficiency. Farrel estimates his production function by fitting an envelope. He has explained the criteria of allocative efficiency and technical efficiency in order to define economic efficiency. In other words to achieve economic efficiency, the farm should be both price efficient and technically efficient. Ainger and Chu used Cobb-Douglas production function to measure technical efficiency. In many agricultural economists Cobb-Douglas production function is the most frequently used production function.

INEFFICIENT USE OF RESOURCES IN AGRICULTURE

Most production economist are not in full accord to the inefficiency in agriculture. They are in general agreement that inefficiency does exist. The causes of inefficiency in the use of agricultural resources are to be found due to gaps between costs and returns of agricultural product. The most important causes of inefficiency of resources-use are lack of knowledge, population growth, unequal changes in demand and supply of factors of agricultural production , regional differences in factor prices etc.

RESOURCES USE EFFICIENCY IN INDIAN AGRICULTURE

The rationality in the allocation of resources shown by the farmers in traditional Indian agriculture was pointed out for the first time by Schultz. Afterwards number of studies have exmined this problem in the traditional and modern agriculture in India. In the past studies, the focus was on the resource-use efficiency at the aggregate crop level. Schultz (1964), A.M. Khusro (1964), Ramkrishna Rao (1965), G.R. Saini

(1969), Sankhayan and Sirohi (1971), Hati and Rudra (1973), J.P. Singh (1975), Bhardwaj and Nadda (1980), Sadhu and Mahajan (1981), P.K. Muraleedharan (1987), Randev, Sharma and Tewari (1990) V.B. Bhise and korna Balakistaiah (1992) G.V. Krishna Rao, Chowdry and Parthasarathy (1994) B.K. Jha (1996), Saswati Das and Chattopadhyay (1998), Uma Devi, Pandurangrao, Shareef and Srinivasulu (2000) etc, studies have examined resource-use efficiency in Indian agriculture.

The development of agricultural sector mainly depends upon the availability of irrigation facilities. Irrigation is an important resource responsible for increasing production of crop. The level of farm income depends on the facilities of water source. So the study of resource-use efficiency on irrigated farm is very important for understanding of increasing agricultural production.

OBJECTIVES

The following are the specific objectives of the study:

1. To study the resource use efficiency in irrigated farms.
2. To study the returns to scale in different crops in the irrigated farms.
3. To asses resource productivity according to size of farms.

METHODOLOGY

Irrigated farms in the command area of Purna Irrigation Project in Maharashtra were selected for the study of resource use efficiency. Data on different aspects such as operational holding, land use pattern, cropping pattern, capital assets, sources of irrigation, family size, infrastructural facilities, total number of beneficiaries etc. were obtained from main office of Purna Irrigation Project.The data on characteristics of the sample farms, inputs and output of the crops in agriculture were obtained with the help of questionnaires by personal interview method.

The sampling design followed was one of stratified random sampling for the selection of villages and farm households. The study area was stratified into three regions based on length of canal form the off take point of the reach namely, head reach, middle reach and tail reach. In the present study, 23 villages (out of 232 villages) were selected from Purna Command Area. A total number of 320 sample farms were studied, out of which 104 sample farms were from head reach, 112 farms from middle reach and 104 farms from tail reach location of command area. From each selected village the sample households were stratified into four size groups based on operational holding viz. marginal (upto 1 hectares), small (1.1to 2.0 hectares), medium (2.1 to 4.0 hectares) and large (above 4 hectares). The total sample was 320 comprising 81 marginal, 97 small, 76 medium and 66 large farms.

The information collected pertained to one complete agricultural year 1999-2000. For the estimation of resource-use efficiency, four foodgrain crops (green gram, kharif jowar, rabi jowar and wheat) and four cash crops (cotton, sugarcane, banana and groundnut) were considered.

FUNCTIONAL ANALYSIS

A functional relationship was developed between output and the input to study the resource use efficiency and returns to scale in selected farms in Purna command area. In the present study, Cobb-Douglas production function was fitted to estimate the resources–use efficiency and returns to scale.All resources of crop production except land input has been measured in value terms and land input measured in hectares.

The Cobb–Douglas production function is specified as follows:

$$Y = a.X_1^{b1}.X_2^{b2}.X_3^{b3}.X_4^{b4}.X_5^{b5}.X_6^{b6}.X_7^{b7}.X_8^{b8}.u$$

Where,

Y = Gross value of output (Rs.)

X_1 = Land area in hectares

X_2 = Value of human labour in rupees

X_3 = Value of bullock labour in rupees

X_4 = Value of manures and fertilizers in rupees

X_5 = Value of pesticides in rupees

X_6 =Value of working capital in rupees.

X_7 = Value of seeds in rupees

X_8 = Irrigation charges in rupees

U = Error term

b_1 to b_8 are the production elasticities of factors X1 to X8 respectively and 'a' is the intercept.

The function in logarithmic scale takes a linear form and is represented as:

$$\text{Log } Y = \text{Log } a + b_1 \log x_1 + b_2 \log x_2 + b_3 \log x_3 + b_4 \log x_4 + b_5 \log x_5 + b_6 \log x_6 + b_7 \log x_7 + b_8 \log x_8 + \log u$$

In the present study output (y) as dependent variable and eight inputs as independent variables were considered to explain the functional analysis. The sum of regression co-efficients (Σbi) indicates returns to scale . The marginal value productivity (MVP) for input x_1 is calculated from the following formula:

$$\text{MVP of } X_1 = \frac{\overline{Y}}{\overline{X}_1} \text{ X } b_1$$

Where,

$\overline{Y}$ = Geometric mean of output

X_1 = Geometric mean of x_1 input

b_1 = Elasticity co efficient of x_1 input

The marginal value productivities for remaining inputs are calculated in the similar manner. The ratio of marginal value product to opportunity cost for each factor was calculated in order to determine the resource–use efficiency. The 't' values were calculated and compared with corresponding table values.

ECONOMIC EFFICIENCY OF RESOURCES

The ratio of marginal value productivities of different resources to their factor costs were calculated to evaluate the efficiency of farmers as users of resources. A ratio, that is equal to unity, optimum use of that factor. A ratio of more than unity indicates that the output could be increased by using that resources and a less than unity ratios indicates the unprofitable level of resources, which should be decreased to minimize the losses.

RESULTS AND DISCUSSION

The resource productivity, returns to scale and resource-use efficiency were estimated for foodgrain (green gram, kharif jowar, rabi jowar and wheat) crops and cash crops (cotton, sugarcane, banana and groundnut) in the command area of Purna irrigation project.

ANALYSIS OF RESOURCE PRODUCTIVITY, RETURNS TO SCALE AND RESOURCE-USE EFFICIENCY ACCORDING TO FARM SIZE

The analysis was made for different size of farms, namely, marginal, small, medium and large and also overall situation i.e. whole sample. The regression Coefficients (Resource Productivity) and related statistics are presented in Table 1.

Resource Productivity

Marginal Farms

It is observed from the Table 1 that the explanatory variables included in the function explained 72 per cent of variation in gross returns from all crops in marginal farms. Among the selected variables the Production elasticities of seeds and irrigation were statistically significant at one per cent level. The regression coefficients of all selected inputs were found to be less than one indicating diminishing factor returns. The regression coefficients of bullock labour and manures and fertilizers were -0.0585 and –0.0038, respectively. These negative coefficients showed that these inputs were excessively used and needs to rationalize these inputs.

Table 1

Analysis of Resource Productivity / Production Elasticities and the Related Statistics for All Crops According to Farm Size in Purna Command Area

Farm size	No. of obser-vations	Degrees of Freedom	Constant	Regression Coefficients / Resource Productivity					Production Elasticities			Returns to Scale (Σbi)	R^2
				Land (Ha.)	Human Labour (Rs.)	Bullock Labour (Rs.)	Manures & Fertili-zers (Rs.)	Value of Pesticides (Rs.)	Working Capital (Rs.)	Value of Seeds (Rs.)	Irrigation Charges (Rs.)		
Marginal	81	72	1.0196	0.0272	0.3488*	-0.0585	-0.0038	0.0001	0.0638	0.3118***	0.4010***	1.0903	0.7252
			(0.1843)	(0.2029)	(0.2565)	(0.2860)	(0.0278)	(0.0160)	(0.0529)	(0.0851)	(0.1376)		
Small	97	88	1.8800	0.1273	0.0145	0.2833**	-0.0014	0.0072	-0.0045	0.3420***	0.1225**	0.8908	0.6469
			(0.1427)	(0.1820)	(0.1676)	(0.1642)	(0.0414)	(0.0130)	(0.0713)	(0.0643)	(0.0652)		
Medium	76	67	1.7169	0.0707	0.1731	0.2014	0.0477**	-0.0105	0.0744	0.0506	0.3203***	0.9277	0.6000
			(0.1107)	(0.1402)	(0.1669)	(0.1948)	(0.0271)	(0.0124)	(0.0745)	(0.0624)	(0.0824)		
Large	66	57	2.5253	0.3973**	-0.4213**	0.3456**	0.1369**	-0.0017	0.0834	0.2946***	0.1597***	0.9945	0.7424
			(0.1289)	(0.2202)	(0.2449)	(0.1912)	(0.0757)	(0.0284)	(0.0786)	(0.0744)	(0.0618)		
Overall	320	311	1.5929	0.1610**	0.0311	0.2756***	0.0158	0.0007	0.0500**	0.2906***	0.1988***	1.0236	0.8779
			(0.1461)	(0.0789)	(0.0975)	(0.0999)	(0.0163)	(0.0070)	(0.0294)	(0.0335)	(0.0393)		

Figures in parentheses re standard errors of the regression coefficients.
*Significant at 10 per cent level.
** Significant at 5 per cent level.
*** Significant at 1 per cent level.

Small Farms

In the case of small farms, the explanatory variables have indicated 64 per cent of variation in gross value of output. The production elasticties of bullock labour, seeds and irrigation were significant. The regression coefficient of manures and fertilizer was –0.0014 and this negative coefficient indicates that the use of manures and fertiuzer was excessive. It was suggested that the use of this resource should be cartailed. The table 1 also showed that production elasticties of all the resources were less then one in the case of small farms indicating diminishing factor returns.

Medium Farms

In case of medium farms, explanatory variables included in the function have indicated 60 per cent of variation in groos value of output. Among the resources selected for the functional analysis, only the production elasticties of manures and fertilizers and irrigation were significant but less than one indicating diminishing factor returns.

Large Farms

The explanatory variables explained 74 per cent of variation in gross value of output in the case of large farms for all crops cultivation in the command area. The production elasticities of land, human labour, bullock labour, manures and fertilizers were significant at 5 per cent level, while the production elasticity of seeds and irrigation were significant at one per cent level, and their positive sign indicates diminishing factor returns. The use of human labour and pesticides were excessive for all crops by the large farms.

Overall Situation

The variables included in the function explained the variation in the gross value of output to the extent of 87 per cent for all crops. Among the selected variables the production elasticites of bullock labour, seeds and irrigation were statistically significant at one per cent level. The study revealed that all the production elasticities were found to be positive and less than one indicating diminishing factor returns.

Returns to Scale

The sum of production elasticities indicates returns to scale. In the case of marginal farms, the sum of production elasticities was 1.0903, the same indicating constant returns to scale since the same was not statistically different from one. The sum of production elastilities (returns to scale) were found to be 0.89, 0.92, 0.99 and 1.023 for small, medium, large and all farms respectively. It was indicated that constant returns to scale were prevailing in cultivation of crops.

Resource-use Efficiency

The ratios of marginal value productivities of different resources to their factor costs (opportunity cost) were calculated to determine the resource–use efficiency. Analysis of geometric mean, marginal value product (MVP), factor costs (opportunity costs) and ratio of MVP to factor costs for all crops are presented in Table 2.

The ratio of MVP to factor costs in the case of marginal farms showed that for human labour (2.4275), working capital (1.3880), seeds (3.83) and irrigation (14.2149) were considerably more than unity indicating that these inputs were under–utilized for all crops production in Purna command area. The ratio was found to be less then one for bullock labour, manures and fertilizers and pesticides indicating that these resources were over–utilized for all crops in the case of marginal farms.

The ratios of marginal value product to factor costs were –0.0143 and –0.1397 for manures and fertilizers and working capital in the case of small farms. Therefore, these resources were used excessively and hence the use of these resources were to be curtailed to get higher returns. The ratio of MVP to factor costs were found negative for pesticides in the case of medium and large farms. Therefor, the use of these resources to be reduced for increasing returns.

In the case of large farms, the ratios of MVP to factor costs for bullock labour (13.2823), manures and fertilizers (6.0968), working capital (11.8485), seeds (11.7309) and irrigation (21.5181) were considerably more than unity indicating that these resources were under-utilized for all crops in the command area.

For the sample as a whole, the ratios of marginal value product to factor costs were 2.8469, 1.5706, 3.4032 and 7.6739 for bullock labour, working capital, seeds and irrigation respectively. These ratios were found to be greater than one indicating that, the expenditure on these resources has to be increased, as this would result in higher production.

REACHWISE ANALYSIS OF RESOURCE PRODUCTIVITY, RETURNS TO SCALE AND RESOURCE-USE EFFICIENCY

The reach wise analysis of resource productivity and returns to scale for all crops cultivation are presented in Table 3.

Resource Productivity

(i) *Head Reach:* The variables included in the function explained the variation in the gross value of output to the extent of 92 per cent on head reach farms. Among these variables, only the resource productivities of seeds and irrigation were turned out to be positive and highly significant at one per cent level of significance and also indicated diminishing factor returns. The resource productivities of

Table 2
Analysis of Geometric mean, Marginal Value Product, Factor Costs (Opportunity Costs) and Ratio of MVP to Factor Cost for All Crops Cultivation According to Farm Size

Sr. No.	Farm Size	Output (Rs.)	Land (Ha.)	Human labour (Rs.)	Bullock Labour (Rs.)	Manures & Fertili-zers (Rs.)	Value of Pesticides (Rs.)	Working Capital (Rs.)	Value of Seeds (Rs.)	Irrigation Charges (Rs.)
1.	Marginal Farm									
	(a) Geometric mean	13539.4181	0.8360	1945.3601	1384.2037	999.5396	22.5424	622.3003	1101.5393	381.9443
	(b) Marginal Value Product	—	440.5169	2.4275	-0.5722	-0.0514	0.0600	1.3880	3.8324	14.2149
	(c) Factor Costs	—	—	1.0000	1.0000	1.0000	1.0000	1.0000	1.0000	1.0000
	(d) Ratio of MVP to Factor Costs	—	—	2.4275	-0.5722	-0.0514	0.0600	1.3880	3.8324	14.2149
2.	Small Farm									
	(a) Geometric mean	31856.6420	1.7963	4384.2973	3088.8726	3112.4329	94.2106	1025.8881	3009.5392	834.2512
	(b) Marginal Value Product	—	2257.6130	0.1053	2.9217	-0.0143	2.4346	-0.1397	3.6201	4.6777
	(c) Factor Costs	—	—	1.0000	1.0000	1.0000	1.0000	1.0000	1.0000	1.0000
	(d) Ratio of MVP to Factor Costs	—	—	0.1053	2.9217	-0.0143	2.4346	-0.1397	3.6201	4.6777
3.	Medium Farm									
	(a) Geometric mean	55475.3435	3.7231	7846.9340	5377.7425	4060.6915	461.1052	1482.8595	4314.1973	1301.9671
	(b) Marginal Value Product	—	1053.4518	1.2237	2.0775	0.6516	-1.2632	2.7833	0.6506	13.6476
	(c) Factor Costs	—	—	1.0000	1.0000	1.0000	1.0000	1.0000	1.0000	1.0000
	(d) Ratio of MVP to Factor Costs	—	—	1.2237	2.0775	0.6516	-1.2632	2.7833	0.6506	13.6476
4.	Large Farm									
	(a) Geometric mean	406443.3292	7.5041	14763.8628	10575.4779	9126.4105	1437.8050	2860.8824	10207.0443	3016.4769
	(b) Marginal Value Product	—	21518.8930	-11.5982	13.2823	6.0968	-0.4805	11.8485	11.7309	21.5181
	(c) Factor Costs	—	—	1.0000	1.0000	1.0000	1.0000	1.0000	1.0000	1.0000
	(d) Ratio of MVP to Factor Costs	—	—	-11.5982	13.2823	6.0968	-0.4805	11.8485	11.7309	21.5181
5.	Overall									
	(a) Geometric mean	38291.2900	2.4490	5265.0100	3706.8000	3104.5500	167.8000	1289.9800	3269.6400	991.9700
	(b) Marginal Value Product	—	2517.3100	0.2261	2.8469	0.1948	0.1597	1.5706	3.4032	7.6739
	(c) Factor Costs	—	—	1.0000	1.0000	1.0000	1.0000	1.0000	1.0000	1.0000
	(d) Ratio of MVP to Factor Costs	—	—	0.2261	2.8469	0.1948	0.1597	1.5706	3.4032	7.6739

Table 3

Reachwise Analysis of Resource Productivity / Production Elasticities and the Related Statistics for All Crops

Reach	*No. of obser-vations*	*Degrees of Freedom*	*Constant (Log a)*	*Regression Productivity (Regression Coefficients*								*Returns to Scale (Σbi)*	R^2
				Land (Ha.)	*Human Labour (Rs.)*	*Bullock Labour (Rs.)*	*Manures & Fertili-zers (Rs.)*	*Value of Pesticides (Rs.)*	*Working Capital (Rs.)*	*Value of Seeds (Rs.)*	*Irrigation Charges (Rs.)*		
Head	104	95	1.7244	0.2172*	0.2416*	0.1719	-0.0074	0.0075	-0.0162	0.2146***	0.1895***	1.0260	0.9244
			(0.1122)	(0.1559)	(0.1540)	(0.1783)	(0.0244)	(0.0120)	(0.0583)	(0.0582)	(0.0674)		
Middle	112	103	0.3763	-0.1300	0.1940	0.3815**	-0.0159	-0.0245**	0.1673***	0.2982***	0.2510***	1.1217	0.8620
			(0.1633)	(0.1782)	(0.2341)	(0.1873)	(0.0835)	(0.0139)	(0.0574)	(0.0659)	(0.1027)		
Tail	104	95	1.0434	0.0274	0.2912**	0.2475**	-0.0037	0.0074	0.0244	0.2922***	0.1260***	1.0124	0.9270
			(0.1185)	(0.1116)	(0.1396)	(0.1496)	(0.0186)	(0.0092)	(0.0413)	(0.0503)	(0.0447)		
Overall	320	311	1.5929	0.1610**	0.0311	0.2756***	0.0158	0.0007	0.500**	0.2906***	0.1988***	1.0236	0.8779
			(0.14161)	(0.0789)	(0.0975)	(0.0999)	(0.0163)	(0.0070)	(0.0294)	(0.0335)	(0.0393)		

Figures in parentheses are standard errors of the regression coefficients.
*Significant at 10 per cent level.
**Significant at 5 per cent level.
***Significant at 1 per cent level.

manures and fertilizers (-0.0074) and working capital (-0.0162) were found to be negative and not significantly contribute to the increase in the value of output at all crops level.

(ii) *Middle Reach:* In the case of middle reach farms, the explanatory variables included in the function have indicated 86 per cent of variation in gross returns.The resource productivities of working capital, seeds and irrigation were 0.1673, 0.2982 and 0.2510 respectively and significant at one per cent level of significance and also indicated the diminishing factor returns. The regression coefficients of land, manures and fertilizer and pesticides were found negative indicating these inputs were used excessively.

(iii) *Tail Reach:* Table 3 reveals that, the explanatory variables included in the function have indicated 92 per cent of variation in gross returns from all crops in tail reach farms of Purna command area. Among the variables selected for the production function, the resource productivities of human labour, bullock labour were positive and significant at 5 per cent level of significance. The regression coefficients of seeds and irrigation inputs were positive, less than unity and significant at 1 per cent level of significance. The regression coefficients of these resources were less than one indicating the operation of diminishing factor returns. The negative resource productivity of manures and fertilizers (-0.0037) indicated the over use of these resources in tail reach farms.

(iv) *Overall:* The resource productivity revealed that, the explanatory variables included in the function have indicated 87 per cent of variation in gross returns for whole sample. The resource productivities of bullock labour (0.2756), seeds (0.2906) and irrigation (0.1988) were found to be highly significant and positive indicating the under use of these resources in all crops cultivation in Purna command area.

Returns to Scale

The sum of production elasticities (returns to scale) of individual resources for all crops in Purna command area was found to be 1.0260 in case of head reach farms and it was tested for its deviation from unity. This indicated the operation of constant returns to scale since it was not statistically different from unity.

In the case of middle reach farms of command area the scale coefficient which is an indicator of the returns to scale was 1.1217. Thought it was more then one, was not statistically different from one indicating the operation of constant returns to scale. The sum of resource productivities of individuals resources for all crops were found to be 1.0124 and 1.0236 for tail reach farm and whole sample respectively.

This sum of elasticities was not statistically different from unity indicating the operation of constant returns to scale.

Resource-use Efficiency

The ratio of marginal value productivities of different inputs to their factor costs were worked out to estimate the resource-use efficiency. Reachwise analysis of geometric mean, marginal value product, factor costs (opportunity costs) and ratio of MVP to factor costs for all crops cultivation are presented in Table 4.

The ratio of marginal value product to factor costs in case of head reach farms were found to be positive and more than one for human labour (1.8437), bullock labour (1.8079), seeds (3.2759) and irrigation (8.0781) indicating the scope for further increase in the usage of these resources to increase the gross income from all crops. For other resources like manures and fertilizers and working capital, it was negative and less than one indicating their excessive use in the crop production.

In the case of middle reach farms, the ratios of marginal value product to factor costs was greater than one for human labour, bullock labour, working capital, seeds and irrigation. This indicated that increase in the use of these resources would bring increase in all crops production in Purna command area. Whereas the ratios were found to be negative for manures and fertilizers and pesticides indicating over use of these resources in the crop production.

The ratios of MVP to factor costs were found to be positive and greater than one for human labour, bullock labour, pesticides, seeds and irrigation in case of tail reach farms indicating the scope for further increase in use of these resources to increase the income from crop production.

For the sample as a whole the profitability from all crops can be increased by increasing the use of bullock labour, working capital, seeds and irrigation and reducing the use of human labour, manures and fertilizers and pesticides.

CONCLUSION

The resource productivity analysis revealed that bullock labour was over utilized in marginal farms and under-utilized in small, medium and large farms. The use of manures and fertilizers in crop production was excessive in the case of marginal and small farms but these resource were under utilized in case of medium and large farms. Irrigation was the most important input for all size of farms for increasing crop production in the command area. The results of production function analysis also indicated that the variables included in the function explained 87 per cent of the variations in gross income.

Table 4
Reach wise Analysis of Geometric mean, Marginal Value Product, Factor Costs (Opportunity Costs) and Ratio of MVP to Factor Cost for All Crops Cultivation.

Sr. No.	*Reach*	*Output (Rs.)*	*Land (Ha.)*	*Human labour (Rs.)*	*Bullock Labour (Rs.)*	*Manures & Fertili-zers (Rs.)*	*Value of Pesticides (Rs.)*	*Working Capital (Rs.)*	*Value of Seeds (Rs.)*	*Irrigation Charges (Rs.)*
1.	Head									
	(a) Geometric mean	42286.33	2.8093	541.15	4188.90	2865.49	375.40	1107.89	2770.12	991.97
	(b) Marginal Value Product	—	3269.35	1.8437	1.8079	-0.1092	0.7447	-0.6183	3.2759	8.0781
	(c) Factor Costs	—	—	1.0000	1.0000	1.0000	1.0000	1.0000	1.0000	1.0000
	(d) Ratio of MVP to Factor Costs	—	—	1.8437	1.8079	-0.1092	0.7447	-0.6183	3.2759	8.0781
2.	Middle									
	(a) Geometric mean	38282.47	2.2289	4356.12	3033.89	4893.27	186.25	1223.48	3543.23	1027.30
	(b) Marginal Value Product	—	-2232.81	1.7049	4.8138	-0.1243	-5.035	5.2347	3.1462	9.3544
	(c) Factor Costs	—	—	1.0000	1.0000	1.0000	1.0000	1.0000	1.0000	1.0000
	(d) Ratio of MVP to Factor Costs	—	—	1.7049	4.8138	-0.1243	-5.035	5.2347	3.1462	9.3544
3.	Tail									
	(a) Geometric mean	34673.68	2.3631	6133.38	4070.05	2060.62	67.05	1335.98	3541.60	954.99
	(b) Marginal Value Product	—	402.03	1.6462	2.1085	-0.0662	3.8267	0.6332	2.8607	4.5747
	(c) Factor Costs	—	—	1.0000	1.0000	1.0000	1.0000	1.0000	1.0000	1.0000
	(d) Ratio of MVP to Factor Costs	—	—	1.6462	2.1085	-0.0662	3.8267	0.6332	2.8607	4.5747
4.	Overall									
	(a) Geometric mean	38291.2900	2.4490	5265.0100	3706.8000	3104.5500	167.8000	1289.9800	3269.6400	991.9700
	(b) Marginal Value Product	—	2517.3100	0.2261	2.8469	0.1948	0.1597	1.5706	3.4032	7.6739
	(c) Factor Costs	—	—	1.0000	1.0000	1.0000	1.0000	1.0000	1.0000	1.0000
	(d) Ratio of MVP to Factor Costs	—	—	0.2261	2.8469	0.1948	0.1597	1.5706	3.4032	7.6739

The results of reach wise analysis indicated that, the resource productivity of land was over utilized in middle reach farms and under utilized in head and tail reach farms. Manures and fertilizers were over utilized for all crop production in the command area. The use of pesticides was excessive in middle reach farms but were under-used on head and tail reach farms.

The results of the resource–use efficiency analysis for marginal farms indicated that there is scope for increasing expenditure on seeds and irrigation, which had statistically significant production elasticities found. The resource productivities of land, bullock labour, manures and fertilizers seeds and irrigation were statistically significant and positive in case of large farms.

The reach wise analysis of resource–use efficiency indicated that, the ratio of MVP to factor costs for human labour (1.8437), bullock labour (1.8079), seeds (3.2759) and irrigation (8.0781) were considerably more than unity in the case of head reach farms indicating that these variables were under–utilized in all crop production in the command area. In the case of middle and tail reach farms the use of manures and fertilizers was found to be excessive. The ratio of MVP to factor costs for irrigation was more than unity and statistically significant indicating that this resource was under–utilized in head, middle and tail reach location of Purna command area.

Lastly, the production function analysis revealed the operation of constant returns to scale and diminishing factor returns in all crops production. The study of marginal value product to factor cost ratios indicated the high degree of resource–use inefficiency in crop production in the command area of Purna irrigation project in Mharashtra.

REFERENCES

Chandrashekar K.S. and Gowda Srinivasa M., Resource-Use Efficiency in Groundnut Production Under Rainfed Condition: A Study in Challakere Taluka of Karnataka, *Agricultural Situation in India,* Vol. 53, No. 6, September, 1996, pp. 387-390.

Koppad M.B., Khan H.S.S. and Shankarmurthy H.G., Resource-use Efficiency in Maize Crop in three locations of Malaprabha Command Area (Karnataka State), *Agricultural Situation in India,* Vol. 54, No. 1 April, 1997, pp. 35-36.

Krishna Rao G.V., Chowdry K.R. and Parthasarathy P.B., Resource Productivity, Returns to Scale and Resource-Use Efficiency in Paddy Farms of Rangareddy District of Andhra Pradesh, *Agricultural Situation in India,* Vol. 49, No. 9, December, 1994, pp. 643-647.

Sampath R.K., Economic Efficiency in Indian Agriculture, Macmillan Company of India Ltd., Mumbai, 1979 p. 64.

Schultz T.W., Transforming Traditional Agriculture, Yale University Press, New Haven, 1964 p.16.

Sikandar Kumar and Sandeep Kumar,Resource Use Efficiency and Returns from Selected Food grain Crops of Himachal Pradesh :A study of Low Hill Zone, *Agricultural Situation in India,* Vol. LXI, No. 7, 2004, pp. 475-484.

Singh J.P., Resource-Use, Farm Size and Returns to Scale in a Backward Agriculture, *Indian Journal of Agricultural Economics,* Vol. 30, No. 2, April-June, 1975, pp. 32-46.

Venkataramana M.N. and Gowda Srinivasa M.V., Productivity and Resource-Use Efficiency in Tomato Cutlivation–An Econometric Analysis, *Agricultural Situation in India,* Vol. 53, No. 6, September, 1986, pp. 409-412.

TAKLE S.R.

BEHAVIOUR OF MARKET ARRIVALS AND PRICES OF AGRICULTURAL COMMODITIES

8

INTRODUCTION

Agricultural production is of a seasonal nature. It is now an established fact that both prices and arrivals of agricultural commodities are subject to variation from one season to another within a year or a week or a month or a period of months which depends upon the duration required for producing the farm commodity and also on its perishability.

The seasonal (temporal or intra-year) variations are up and down swings in prices occur with some regularity during the year. The seasonal variations resemble a cycle occupying a period of 12 months or less, due to low capacity of the small and marginal farmers to hold stock and to obtain credit to postpone their sales, less storability of agricultural products etc.

One of the most important aspects of agricultural prices is the wide range of seasonal variations. The reason behind it is that the repayment of loans taken from money lenders, is made soonafter the harvesting of crops. Hence the farmers are forced to sell their produce immediately, after the harvest. Such sales are called 'distress sales'. Acharya and Agrawal (1994) say that price variations are observed for agricultural commodities mainly because of the nature of agricultural production.

The analysis of the longterm movements in prices helps in ascertaining as to what extent the marketing system is efficient and also helps in production planning. Further, the behaviour of prices over a period of time helps to understand the price structure and movements. It is necessary to study the farm prices which forms an essential requirement for policy formulation. On the basis of price studies the cultivators can make decisions for proper allocation of crop areas by anticipating future prices based on the prices which prevented in the past. The knowledge of structure and movement of prices and arrivals and of their relationship is essential for enabling the farmers to reap the full benefit of their harvests.

In this paper an attempt is made to study the seasonal variation in prices and market arrivals of commodities at regulated markets in Marathwada, and the trends in prices and arrivals of agricultural commodities at regulated markets in Marathwada.

DATA AND METHODOLOGY

Study Area

Marathwada is a backward region of the Maharashtra state. The main occupation of people in this region is agriculture. At present 75 per cent of the total population depends on agriculture in this region. If agricultural sector is developed it would be possible to achieve the development of industrial sector as well as the whole economy of the Marathwada region.

Selection of Markets and Commodities

This study covers seven districts of this region. One marketing centre from each district of the Marathwada region is chosen for the present study. In all 7 regulated markets are selected for the purpose of the present study, which are Aurangabad, Jalna, Beed, Osmanabad, Nanded, Parbhani & Latur. This study covers six main agricultural commodities produced in this region. The commodities are wheat, Kharif jowar, rabi jowar, tur, safflower and groundnut.

After having collected the data on monthly arrivals and prices of agricultural commodities from the office records of the chosen regulated markets, we came to know that in case of kharif jowar, safflower and groundnut arrivals are low. Hence we decided to drop such markets. All the seven markets have been considered in case of wheat, rabi jowar and tur.

This study covers the period of 15 years, from 1981 to 1995, and is based completely on secondary data. The time series data regarding monthly arrivals and average prices of the commodities are collected from the records of the agricultural produce market committees in Marathwada.

Tools of Analysis

In the study some statistical tools have been used for analysing the data on arrivals and prices of agricultural commodities. These tools are as below:

Seasonal variation

The variations in the monthly market arrivals and monthly average prices of agricultural commodities are measured by using the method of simple averages.

(i) Range: In order to measure the seasonal variations in arrivals and prices, the range between the maximum and minimum values of arrivals and prices of agricultural commodities is taken.

$$R_0 = R_b - R_a$$

Where,

R_0 = Range

R_b = Maximum level

R_a = Minimum level

(ii) Standard Deviation (S.D.): The technique of standard deviation is employed to measure the seasonal variations in arrivals and prices of agricultural commodities. The standard deviation (S.D.) is calculated as below.

$$\sigma = \sqrt{\frac{\sum dx^2}{N}}$$

Where,

σ = Standard deviation

dx = deviation from actual mean i.e. $(X - \overline{X})$

$\overline{X}$ = arithmetic mean

n = Total number of observations.

(iii) Coefficient of Variation (C.V.): Coefficient of variation is a relative measure of dispersion or variation of market arrivals and average prices of agricultural commodities. It is calculated as below:

$$C.V. = \frac{\sigma}{\overline{X}} \times 100$$

Where,

C.V. = Coefficient of variation,

σ = standard deviation, and

X = arithmetic mean.

TREND

The trend of arrivals of each agricultural commodity is studied by fitting the linear regression equation of arrivals on time with the help of least-squares method. Similarly, the trend of prices of each agricultural commodity is studied by fitting the linear regression equation of prices on time with the help of least-squares method. The simple linear regression equation used to estimate the annual trend values of arrivals and prices of each agricultural commodity for each regulated market is as follows:

$$Y = a + bX$$

Where,

Y = Arrivals/prices,

X = Time (in years),

a = Intercept, and

b = Slope.

Range, standard deviation and coefficient of variation of monthly arrivals and average prices have been calculated separately for each commodity and market considered in the present study.

RESULTS AND DISCUSSION

Seasonal Variation

The seasonal nature of agricultural production, low capacity of the small and marginal farmers to hold stocks of farm product and to obtain credit to postpone their sales and less storability of agricultural products are the factors which explain the variation in arrivals and prices of agricultural commodities from one season to another within a year. The range, standard deviation, coefficient of variation of monthly arrivals and average prices have been calculated separately for each commodity and market considered in the present.

Wheat: Wheat is the most important crop grown in India, as well as in Marathwada. The results obtained after analysing the data on monthly arrivals and average prices of wheat at regulated markets in Marathwada region are presented in table 1. This table shows that the range between the monthly maximum and minimum arrivals of wheat is the highest in Jalna (7679.46 quintals), and the lowest range in arrivals is found in Osmanabad (436.67 quintals).

Both standard deviation and coefficient of variation in arrivals of wheat are found to be the highest in Jalna (2372.72) and Nanded (70.53) markets and the lowest in Osmanabad (145.16) and Aurangabad (54.09) markets, because the size of wheat arrivals is relatively higher in Jalna and Nanded markets.

As far as the monthly variation in prices is concerned, the value of standard deviation and coefficient of variation are found to be highest in case of Jalna market (i.e. 19.58 and 5.59 respectively) and the lowest in Latur (i.e. 12.57 and 3.58 respectively) market. Because in Jalna market arrival of wheat is low as compared to Latur market and therefore the price fluctuations are different. Further, the data presented in table 1 reveals that the degree of variations in monthly arrivals is higher than that in monthly prices in case of wheat. It means that the arrivals of wheat have shown greater monthly variations than its prices.

Table 1

Seasonal Arrivals and Average Prices of Wheat at Regulated Markets in Marathwada During 1981 to 1995

(Arrivals in quintals)
(Price in Rs. per quintal)

Months	*Markets*													
	Aurangabad		*Latur*		*Jalna*		*Osmanabad*		*Parbhani*		*Nanded*		*Beed*	
	Arrivals	*Prices*	*Arrivals*	*Prices*	*Arrivals*	*Prices*	*Arrivals*	*Prices*	*Arrivals*	*Prices*	*Arrivals*	*Prices*	*Arrivals*	*Prices*
January	1418.33[a]	352.20	1291.87	363.20	2458.07	360.40	120.27	364.53	475.33	326.67	848.93	334.73	657.20[a]	358.40
February	1816.07	340.73	2188.47	343.87	2839.73	338.73	215.73	345.27	622.60	337.13	689.87[a]	329.33	798.47	374.33
March	5525.73[b]	311.87	5633.80[b]	332.67[a]	8002.53	316.20[a]	523.33	328.00[a]	1172.07	320.13	3821.80	293.40[a]	2632.73	326.60[a]
April	5316.40	307.07	5099.67	340.07	9634.53[b]	336.20	545.20[b]	351.40	1172.47[b]	328.40	3288.53	302.60	3212.93[b]	339.00
May	3510.33	303.80[a]	2659.93	345.33	5106.20	328.07	365.60	350.00	632.80	335.33	1357.20	303.53	2012.07	338.87
June	2840.74	308.07	1837.27	333.47	4039.60	325.67	276.27	346.60	496.07	333.87	1119.80	306.93	1479.67	344.00
July	2083.93	325.60	1254.00	349.67	2920.40	331.93	160.87	363.47	381.33	352.67	994.87	323.20	1062.27	353.33
August	2362.73	327.33	1425.47	354.00	2893.13	335.93	172.27	363.53	400.87	347.87	840.53	330.20	999.07	358.80
September	2116.40	330.47	1111.73[a]	352.33	1955.07[a]	344.60	143.80	366.27	311.27[a]	315.87[a]	4109.67[b]	328.00	920.33	359.73
October	2014.73	329.13	1134.40	357.00	2282.93	356.20	227.80	374.47	371.73	330.60	917.87	325.93	932.93	365.83
November	1918.27	339.27	1478.07	365.33	2282.93	365.27	135.20	378.40	420.87	361.67	1416.07	337.53	693.13	377.73
December	1571.53	368.20[b]	1459.33	376.47[b]	2424.00	389.47[b]	108.53[a]	386.27[b]	369.20	389.33[b]	1062.80	351.80[b]	678.40	395.40[b]
Average	2707.91	328.65	2214.50	351.12	3903.26	344.06	249.57	359.85	568.88	339.96	1705.66	322.27	1339.93	357.63
Range between Maximum and Minimum	4107.40	64.40	4522.07	43.80	7679.46	73.27	436.67	58.27	861.20	73.46	3419.80	58.40	2555.73	68.80
S.D.	1464.63	18.66	1478.99	12.57	2372.72	19.58	145.16	15.59	285.71	19.52	1202.94	16.40	806.28	18.35
C.V.	54.09	5.68	66.79	3.58	60.79	5.69	58.16	4.33	50.22	5.74	70.53	5.09	60.17	5.13

a = Minimum level. b = Maximum level S.D. = Standard deviation C.V. = Coefficient of variation.

Source: The data on monthly arrivals and average prices are compiled from the office records of regulated markets.

Kharif Jowar: Kharif Jowar is the main cereal crop of the Marathwada region. The results presented in the table 2 indicate that the range between the monthly maximum and minimum arrivals of kharif jowar is the highest in Latur (28508.34 quintal) market and the lowest in Beed (1311.34 quintal) market. The range in monthly prices of jowar is found to be the highest in Parbhani (Rs. 56.20) market and lowest in Aurangabad (Rs. 31.27) market. It means that the arrivals of kharif jowar have been subject to greater variation in Latur and lesser variation in Beed market and the degree of monthly variation in prices is found to be the highest in Parbhani market and the lowest in Aurangabad market. From this it is clear that the arrivals and prices of Kharif jowar have varied in different degrees in different markets.

The standard deviation and coefficient of variation of arrivals of Kharif jowar are found to be the highest in Latur (9368.32) market and Parbhani (96.62) market and the lowest in Beed (447.18) market and Latur (59.99) market. As far as monthly variation in prices is concerned it is found that the value of standard deviation and coefficient of variation are the highest for Parbhani (i.e. 13.31 and 7.64 respectively) market and the lowest for Osmanabad (8.74 and 4.43 respectively) market. It means that the month to month variations in prices of Kharif jowar have been relatively higher in Parbhani market and relatively lower in Osmanabad market. The degree of seasonal variation of prices of kharif jowar found in case of Latur market may be due to low production.

Rabi Jowar: Rabi jowar is an important crop of the Marathwada region. In Maharashtra state, rabi jowar is grown mainly in Marathwada region. The results presented in table 3 show that the range between monthly maximum and minimum arrivals is the highest in Jalna (13448.20 quintals) market and the lowest in Osmanabad (615.13 quintals) market. In the case of monthly prices of rabi jowar the range between the maximum and minimum prices is found to be the highest in Parbhani (Rs. 87.86) market and the lowest in Nanded (Rs. 33.80) market.

The standard deviation and coefficient of variation in arrivals is found to be the highest in Jalna (i.e. 4399.60, 55.28) market. On the other hand, the lowest value of coefficient of variation is found for Parbhani (29.10) market. But an interesting thing to be noted here is that the standard deviation and coefficient of variation in monthly prices of rabi jowar are found to be the highest in Parbhani (i.e. 23.54 and 7.42 respectively) market and the lowest in Nanded (8.70 and 4.11 respectively) market. From this it can be said that prices of rabi jowar have been subject to lesser variation in Nanded market during the period of this study. And the prices have varied to a larger extent in Parbhani market.

Table 2
Seasonal Arrivals and Average Prices of Kharif Jowar at Regulated Markets in Marathwada During 1981 to 1995

(Arrivals in quintals), (Price in Rs. per quintal)

Months	*Markets*											
	Aurangabad		*Latur*		*Jalna*		*Osmanabad*		*Parbhani*		*Beed*	
	Arrivals	Prices	Arrivals	Prices	Arrivals	Prices	Arrivals	Prices	Arrivals	Prices	Arrivals	Prices
January	2179.73	182.93	25826.67	191.40	6018.87	195.67	1503.20	190.87	1050.13	180.27	1224.13	201.73
February	1744.53	185.27	16076.53	194.33	4672.47	196.87	843.27	197.60	1100.13	185.07	656.40	196.80
March	1082.13	186.00	12976.27	191.20	2755.60	192.53	427.80	198.33	541.27	161.93	345.53	202.20
April	1261.00	208.67	12874.27	199.07	2895.87	196.87	313.40	197.27	337.47	175.75	318.93	209.60
May	911.13	194.13	8866.00	203.13	2782.93	201.67	278.00	197.87	145.40	168.33	256.07	213.53
June	1478.93	195.67	9662.60	206.93	2810.67	209.73	257.07	199.00	392.67	177.33	214.80	202.07
July	1259.33	214.07[b]	7412.27	216.60[b]	2556.20	217.13[b]	274.47	199.93	91.87[a]	179.47	207.07	221.87
August	1346.47	202.73	8413.67	212.87	2686.13	202.60	298.87	209.27	153.20	139.67[a]	288.87	225.73[b]
September	775.93[a]	198.27	6339.93[a]	200.80	1613.13[a]	206.20	209.47[a]	214.47[b]	162.80	195.87[b]	167.13[a]	207.53
October	6082.40	182.80[a]	12159.67	184.00[a]	10987.20	179.33[a]	884.47	177.67[a]	1162.27	170.27	956.47	192.27[a]
November	6849.47[b]	188.20	31930.93	188.80	15857.33[b]	185.93	2352.60[b]	191.27	2781.60[b]	175.13	1478.47[b]	195.93
December	3401.60	185.47	34848.27[b]	203.33	7308.93	203.07	2148.13	195.40	1465.67	182.53	1109.07	206.33
Average	2364.39	193.68	15615.59	199.37	5245.44	198.97	815.90	197.41	782.04	174.30	601.91	206.30
Range between maximum and minimum	6073.54	31.27	28508.34	32.60	14244.20	37.80	2143.13	36.80	2689.73	56.20	1311.34	33.46
S.D.	1983.24	10.04	9368.32	9.45	4097.28	9.80	738.64	8.74	755.59	13.31	447.18	9.72
C.V.	83.88	5.18	59.99	4.74	78.11	4.92	90.53	4.43	96.62	7.64	74.29	4.71

a = Minimum level. b = Maximum level S.D. = Standard deviation C.V. = Coefficient of variation.
Source: As indicated in table 1

Table 3
Seasonal Arrivals and Average Prices of Rabi Jowar at Regulated Markets in Marathwada During 1981 to 1995.

(Arrivals in quintals)
(Price in Rs. per quintal)

Months	*Markets*													
	Aurangabad		*Latur*		*Jalna*		*Osmanabad*		*Parbhani*		*Nanded*		*Beed*	
	Arrivals	*Prices*	*Arrivals*	*Prices*	*Arrivals*	*Prices*	*Arrivals*	*Prices*	*Arrivals*	*Prices*	*Arrivals*	*Prices*	*Arrivals*	*Prices*
January	1876.07	312.00	1052.00	333.33	3746.27[a]	307.80	271.20[a]	304.87	533.53	335.60	1853.53	208.73	2000.67[a]	303.93
February	1703.73	302.33	900.20[a]	317.80	4596.73	293.80[a]	288.47	299.20	793.27	289.00	1151.47[a]	206.13	2076.47	296.53
March	3464.73	297.33[a]	1697.93	317.33	15627.20	295.27	505.53	290.53[a]	851.27	304.07	1314.60	208.47	5385.93	294.47[a]
April	3961.87[b]	309.13	3187.87[b]	328.47	17194.47[b]	304.13	886.33[b]	314.20	1255.13[b]	275.07[a]	1582.93	214.47	7067.40[b]	305.13
May	2639.80	305.13	1892.87	334.33	10728.13	297.20	660.00	304.47	764.20	313.33	1332.00	211.20	5326.00	296.53
June	2294.27	307.27	2121.13	320.67	9324.33	298.67	513.60	312.33	659.40	310.20	1730.33	214.00	4731.27	297.67
July	2017.60	308.53	1630.33	324.80	8538.40	313.07	406.33	326.20	592.87	312.67	2018.73	228.33[b]	4654.47	306.60
August	1844.13	314.20	1418.13	313.13[a]	7892.13	297.40	532.33	306.67	703.07	362.93[b]	2166.60	216.27	4116.47	304.00
September	1881.47	309.87	1753.13	324.87	4829.87	308.60	530.07	313.27	614.53	319.93	5906.07[b]	222.80	3557.20	299.07
October	1470.47	310.87	1527.20	330.67	4688.80	309.40	424.73	319.73	649.53	309.20	2027.13	194.53[a]	2846.27	314.60
November	1022.47[a]	331.07	1359.20	361.33	4150.27	333.47	306.20	350.47	482.60	321.33	2987.80	200.47	2277.47	339.13
December	1230.53	348.27[b]	1221.40	363.67[b]	4182.00	345.33[b]	334.13	358.67[b]	470.60[a]	353.33	2452.60	213.73	2111.13	340.47[b]
Average	2117.26	313.00	1646.78	330.87	7958.22	308.68	471.58	316.72	697.50	317.22	2210.32	211.59	3845.90	308.18
Range between maximum and minimum	2939.40	50.94	2287.67	50.54	13448.20	51.53	615.13	68.14	784.53	87.86	4754.60	33.80	5066.73	46.00
S.D.	977.95	13.15	571.26	15.47	4399.60	15.16	169.29	19.19	202.97	23.54	1219.29	8.70	1568.95	15.12
C.V.	46.19	4.20	34.69	4.68	55.28	4.91	35.90	6.06	29.10	7.42	55.16	4.11	40.80	4.90

a = Minimum level. b = Maximum level S.D. = Standard deviation C.V. = Coefficient of variation.
Source: As indicated in table 1.

Tur: Tur is the important pulse crop in Marathwada. In case of tur table 4 reveals that the range between the monthly maximum and minimum arrivals is the highest in Latur (32719.40 quintals) market and the lowest in Osmanabad (2386.07 quintals) market. But the price range is found to be the highest in Aurangabad (Rs. 154.20) market and the lowest in Jalna (Rs. 114.33) market.

The standard derivation and coefficient of variation in monthly arrivals of tur are found to be the highest in Latur (10282.48) and Aurangabad (136.49) markets respectively and the lowest in Osmanabad (731.49) and Nanded (80.30) markets respectively. It means that the arrivals of this commodity are subject to relatively greater variation in Latur and Aurangabad markets and relatively lesser variation in arrivals in Osmanabad and Nanded markets. As far as monthly variation in prices of tur is concerned, both the standard deviation and coefficient of variation are the highest for Aurangabad (i.e. 46.11 and 6.76 respectively) market and the lowest for Jalna (i.e. 38.28 and 5.29 respectively) market. This indicates that the prices of tur commodity have been subject to lesser variation in Jalna market and have varied to a greater extent in Aurangabad market. The lowest price variation in Aurangabad market may be due to low level of arrivals of this commodity.

Safflower: Safflower is an important oilseed crop grown in rabi season on medium to black soils in the Marathwada region. The results presented in table 5 reveal that the range between the monthly maximum and minimum arrivals of Safflower is the highest in Jalna (6548.46 quintals) market and the lowest in Osmanabad (212.67 quintals) market. As far as monthly prices are concerned, the range between maximum and minimum is found to be the highest in Parbhani (Rs. 281.47) market and the lowest in Latur (Rs. 136.73) market.

The standard deviation and coefficient of variation in monthly arrivals is found to be the highest for Jalna (1959.79) and Osmanabad (28.69) market and the lowest for Osmanabad (60.97) and Jalna (64.12) market. It means that the variations in monthly arrivals have been of smaller degree in Jalna market and of greater degree in Osmanabad market during the period of this study. But both standard deviation and coefficient of variation in monthly prices of Safflower are found to be the highest in Parbhani (i.e. 85.68 and 14.65 respectively) market and the lowest in Aurangabad (i.e. 43.05 and 6.45 respectively) market.

Groundnut: Groundnut is an important oilseed crop grown in Marathwada (See table 6). In case of groundnut the range between the monthly maximum and minimum levels of arrivals is found to be the highest in Nanded (13375.33 quintals) market and the lowest in Osmanabad (337.26 quintals) market. But the range of its monthly prices is found to be the highest in Jalna (Rs. 398.87) market and the lowest in Manjlegaon (Rs.100.80) market.

Table 4

Seasonal Arrivals and Average Prices of Tur at Regulated Markets in Marathwada During 1981 to 1995.

(Arrivals in quintals)
(Price in Rs. per quintal)

Months	*Markets*													
	Aurangabad		*Latur*		*Jalna*		*Osmanabad*		*Parbhani*		*Nanded*		*Beed*	
	Arrivals	*Prices*	*Arrivals*	*Prices*	*Arrivals*	*Prices*	*Arrivals*	*Prices*	*Arrivals*	*Prices*	*Arrivals*	*Prices*	*Arrivals*	*Prices*
January	9727.93[b]	670.40	34855.87[b]	690.40	25605.20[b]	676.87	2386.07[a]	670.13	8018.40[b]	667.33	3820.67[b]	654.53	9697.67[b]	671.20
February	4013.33	649.73	26763.73	686.33[a]	15428.13	671.00[a]	1882.20	658.00	5262.60	635.80[a]	3747.67	650.33[a]	5729.93	669.87[a]
March	1820.53	663.20	20764.07	700.93	8689.33	678.27	1032.20	655.93[a]	3995.53	638.53	3783.27	656.53	3229.73	676.33
April	1370.93	664.80	19087.87	707.73	6571.73	684.47	982.53	679.33	2341.20	645.47	2248.73	671.27	2095.07	684.93
May	596.20	639.80	14259.40	729.00	3436.13	699.53	531.27	682.13	1746.47	656.13	1711.13	681.93	931.00	702.13
June	559.40	694.20	10474.27	768.00	2509.73	734.20	398.33	755.13	1195.73	706.80	1190.60	726.60	666.00	760.13
July	222.40	748.27	6217.00	808.40[b]	1743.40	767.53	196.73	781.07[b]	771.40	738.53	890.33	759.53	409.47	790.33[b]
August	132.93	660.67	3395.47	787.33	1030.47	738.00	116.47	727.73	271.27	741.73	531.53	741.13	172.67	755.40
September	385.00	624.20[a]	2137.00	784.53	742.00	756.87	63.07[a]	770.27	226.93	680.33	609.93	746.60	71.07	734.67
October	98.13[a]	648.60	2136.47[a]	772.33	585.73[a]	756.67	82.80	712.27	254.60	719.20	293.80	766.93[b]	64.87[a]	717.27
November	538.47	738.27	2166.20	776.33	815.73	738.00	117.13	713.27	211.40[a]	683.93	235.40[a]	735.53	299.27	731.87
December	9116.73	778.40[b]	10723.07	805.67	2323.80	785.33[b]	1129.00	749.20	3314.40	755.47[b]	936.73	738.40	3963.73	768.67
Average	2381.83	681.71	12748.37	751.42	5790.12	723.90	743.65	712.87	2300.83	689.10	1666.65	710.78	2277.54	721.90
Range between maximum and minimum	9629.80	154.20	32719.40	122.07	25019.47	114.33	2386.07	125.14	7807.00	119.67	3585.27	116.60	9632.80	120.46
S.D.	3250.92	46.11	10281.46	43.62	7298.44	38.28	731.49	42.25	2348.84	40.62	1338.27	42.34	2839.73	39.63
C.V.	136.49	6.76	80.65	5.80	126.05	5.29	98.37	5.93	102.09	5.90	80.30	5.96	124.68	5.49

a = Minimum level. b = Maximum level S.D. = Standard deviation C.V. = Coefficient of variation.
Source: As indicated in table 1.

Table 5
Seasonal Arrivals and Average Prices of Safflower at Regulated Markets in Marathwada During 1981 to 1995

(Arrivals in quintals)
(Price in Rs. per quintal)

Months	*Markets*											
	Aurangabad		*Latur*		*Jalna*		*Osmanabad*		*Parbhani*		*Beed*	
	Arrivals	*Prices*	*Arrivals*	*Prices*	*Arrivals*	*Prices*	*Arrivals*	*Prices*	*Arrivals*	*Prices*	*Arrivals*	*Prices*
January	188.67[a]	661.07	246.13[a]	628.60	1955.80	651.80	4.40	463.87	126.87	546.80	71.47[a]	460.07[a]
February	1483.13	661.33	1281.40	557.47[a]	4546.13	589.93[a]	76.27	496.73	423.07	452.47	998.40	525.60
March	3228.40[b]	590.27[a]	2504.87[b]	561.47	7704.13[b]	593.93	215.20[b]	571.47	2358.53[b]	580.60	3237.80[b]	589.87
April	1882.33	606.93	1800.00	590.53	5395.33	613.13	121.27	587.40	2189.87	608.60	2002.93	617.53
May	871.00	614.80	877.67	600.87	3090.40	637.73	51.33	599.93	946.53	634.00	1237.13	631.93
June	618.93	646.60	761.20	639.73	2907.93	668.00	34.13	629.60[b]	517.07	652.93	968.47	657.13
July	328.27	684.80	451.27	681.20	2662.53	711.73	14.67	549.20	317.40	687.07	640.40	617.13
August	272.73	703.27	363.13	689.47	1792.93	699.67	16.27	608.40	168.53	555.93	455.73	634.20
September	220.27	688.80	334.13	674.40	1560.27	693.60	10.47	585.53	70.73[a]	407.33[a]	322.80	613.40
October	267.93	701.20	452.93	665.13	1526.13	705.93	16.33	575.00	132.47	540.40	370.40	694.60[b]
November	322.33	716.27	371.27	683.80	1155.67[a]	729.07	5.67	584.00	124.93	688.80[b]	153.07	635.67
December	346.47	728.93[b]	260.40	694.20[b]	1307.00	739.33[b]	2.53[a]	424.20[a]	84.27	661.67	82.27	625.60
Average	835.87	667.02	808.70	638.91	2967.02	669.49	47.38	556.28	621.69	584.72	878.41	608.56
Range between maximum and minimum	3039.73	138.66	'2258.74	136.73	6548.46	149.40	212.67	205.40	2287.80	281.47	3166.33	234.53
S.D.	882.73	43.05	681.71	48.21	1902.51	49.64	60.97	59.69	777.44	85.68	894.22	58.77
C.V.	105.61	6.45	84.30	7.55	64.12	7.41	128.69	10.73	125.05	14.65	101.80	9.66

a = Minimum level. b = Maximum level S.D. = Standard deviation C.V. = Coefficient of variation.
Source: As indicated in table 1.

Table 6
Seasonal Arrivals and Average Prices of Groundnut at Regulated Markets in Marathwada During 1981 to 1995

(Arrivals in quintals)
(Price in Rs. per quintal)

Months	*Markets*											
	Jalna		*Parbhani*		*Latur*		*Beed*		*Nanded*		*Osmanabad*	
	Arrivals	*Prices*	*Arrivals*	*Prices*	*Arrivals*	*Prices*	*Arrivals*	*Prices*	*Arrivals*	*Prices*	*Arrivals*	*Prices*
January	207.27	506.47	74.07	660.33	4616.60	723.73	2189.33	736.73	570.53	705.60	247.73	704.80
February	48.07	558.13	50.13	527.13	4659.53	758.00	995.13	744.67	177.60	532.47	142.73	735.53
March	10.00[a]	332.13	20.93	476.80	2450.20	778.13	629.47	742.20	110.00	516.60[a]	98.27[a]	766.73
April	10.80	331.33[a]	13.87[a]	430.80[a]	2932.00	788.73	497.40	755.20	80.87[a]	620.40	107.53	782.67[b]
May	50.60	596.00	1286.00	644.13	4035.93	795.00	289.53[a]	721.47[a]	4354.53	751.33	101.93	759.87
June	240.80	695.93	2767.81[b]	733.27	5780.00[b]	800.33	487.13	751.13	13456.20[b]	779.00	172.47	742.27
July	435.20	727.27	1962.07	816.47[b]	4526.87	833.33[b]	448.60	826.00[b]	5328.87	849.67	134.93	715.33
August	548.33	705.53	401.60	697.13	3498.60	799.07	656.20	758.13	1132.87	874.53[b]	278.13	708.47
September	676.67	652.67	221.07	578.67	3183.07	767.87	2127.47	733.87	795.13	780.33	435.53[b]	636.93[a]
October	889.27	637.93	142.87	581.87	2727.00	766.40	3057.60	737.80	685.53	797.33	303.13	683.53
November	1128.53[b]	707.20	162.67	732.07	3522.07	710.67[a]	3612.00[b]	721.73	809.20	767.73	409.20	719.27
December	685.80	730.20[b]	114.73	734.27	666.33[a]	737.33	3377.27	761.93	687.67	768.93	400.27	728.53
Average	410.95	600.190	601.49	634.41	3549.85	771.55	1530.59	749.24	2349.08	728.66	235.99	723.66
Range between maxi.and min.	1118.53	398.87	2753.94	385.67	5113.67	122.66	3322.47	104.53	13375.33	357.93	337.26	145.74
S.D.	360.81	136.02	870.90	112.10	1266.70	33.76	1210.89	26.29	3720.43	110.01	122.34	37.60
C.V.	87.80	22.64	144.79	17.67	35.68	4.38	79.11	3.51	158.38	15.10	51.84	5.20

a = Minimum level. b = Maximum level S.D. = Standard deviation C.V. = Coefficient of variation.
Source: As indicated in table 1.

The standard deviation and coefficient of variation in monthly arrivals of this commodity is found to be the highest in Nanded (i.e. 3720.43, 158.38 respectively) market and the lowest in Osmanabad (122.34) and Latur (35.68) markets. This indicates that the arrivals of groundnut have been subject to greater variation in Nanded market during the period under study. Because in Nanded market area, groundnut is taken during summer season on irrigated lands. So the production of groundnut in this area depends largely on the availability of irrigation water. It is also found that the variations in monthly arrivals have been of smaller degree in Latur and Osmanabad markets. Both standard deviation and coefficient of variation in monthly prices of groundnut are found to be the highest in Jalna (i.e. 136.02 and 22.45 respectively) market and the lowest in Beed (i.e. 26.29 and 3.51 respectively) market. In case of Jalna market the degree of variation in monthly price is higher because the production of groundnut is lower in this market area. The data also indicates that both arrivals and prices of groundnut have varied in different degrees in different markets during the period of this study.

TREND

Trend is the basic tendency of a series of growth or decline over a period of time. It measures the changes either up or down, of the values over a long period of time. The trend is that component of variation which reveals the general direction of change over a period of time. Trend is usually established on the basis of at least 10-15 years data. The concept of trend does not include short range oscillations but rather steady movements over a long time. Under trend, changes (increases or decreases) are not sudden but are of slow moving tendencies. This component in price variation is affected by factors such as changes in demand arising from changes in population, income, dietary habits, customs and establishment of processing industries for the commodity concerned. Price trend is also affected by adjustment in supply arising out of development of cold storage and marketing facilities, production technology and market arrivals. Thus, the long term smooth movements are the result of the forces which change gradually over a long period of time. As a result, we can notice a number of factors including the changes in prices and the variation in arrivals.

After looking at the aspect of trend in general, agricultural commodity-wise trend in arrival and prices is given below:

Wheat: The annual trend values of arrivals and prices of wheat for regulated markets are given in table 7. It is observed from this table that annual trend values of arrivals of wheat have shown a rising trend over the period in four regulated markets viz., Aurangabad, Jalna, Osmanabad and Nanded. Among the markets where annual trend values

Table 7
Trend Values of Arrivals and Prices of Wheat at Regulated Markets in Marathwada During 1981-95

(Arrivals in '00 Quintals)
(Price in Rs. per Quintal)

Years	*Markets*													
	Aurangabad		*Parabhani*		*Jalna*		*Nanded*		*Latur*		*Beed*		*Osmanabad*	
	Trends values of arrivals	*Trends values of prices*	*Trends values of arrivals*	*Trends values of prices*	*Trends values of arrivals*	*Trends values of prices*	*Trends values of arrivals*	*Trends values of prices*	*Trends values of arrivals*	*Trends values of prices*	*Trends values of arrivals*	*Trends values of prices*	*Trends values of arrivals*	*Trends values of prices*
1981	191.09	174.26	127.95	180.37	413.80	171.10	172.25	164.31	398.68	164.11	164.16	164.29	23.56	173.96
1982	210.18	196.42	119.36	203.17	421.72	195.81	176.88	186.88	379.69	190.82	163.67	191.91	24.46	200.52
1983	229.27	218.58	110.77	225.97	429.56	220.52	181.51	209.45	360.70	217.53	163.18	219.53	25.36	227.08
1984	248.36	240.74	102.18	248.77	437.40	245.23	186.14	232.02	341.71	244.24	162.69	247.15	26.26	253.64
1985	267.45	262.90	93.59	271.57	445.24	269.94	190.77	254.59	322.72	270.95	162.20	274.77	27.16	280.20
1986	286.54	285.06	85.00	294.37	453.08	294.65	195.40	277.16	303.73	297.66	161.71	302.39	28.06	306.76
1987	305.63	307.22	76.41	317.17	460.92	319.36	200.03	299.73	284.74	324.37	161.22	330.01	28.96	333.32
1988	324.72	329.38	67.82	339.97	468.76	344.07	204.66	322.30	265.75	351.08	160.73	357.63	29.86	359.88
1989	343.81	351.54	59.23	362.77	476.60	368.78	209.29	344.87	246.76	377.79	160.24	385.25	30.76	386.44
1990	362.90	373.70	50.64	385.57	484.44	393.49	213.92	367.44	227.77	404.50	159.75	412.87	31.66	413.00
1991	381.99	395.86	42.05	408.37	492.28	418.20	218.55	390.01	208.78	431.21	159.26	440.49	32.56	439.56
1992	401.08	418.02	33.46	431.17	500.12	442.91	223.18	412.58	189.79	457.92	158.77	468.11	33.46	466.12
1993	420.17	440.18	24.87	453.97	507.96	467.62	227.81	435.15	170.80	484.63	158.28	495.73	34.36	492.68
1994	439.26	462.34	16.28	476.77	515.80	492.33	232.44	457.72	151.81	511.34	157.79	523.35	35.26	519.24
1995	458.35	484.50	7.69	499.57	523.64	517.04	237.07	480.29	132.82	538.05	157.30	550.97	36.16	545.80

Note: Trend values of arrivals and prices are computed by employing the least-squares method.
Source: Data on yearly arrivals and prices are compiled from the office records of the regulated markets.

of arrivals showed rising trend, the highest base year value of arrivals (i.e. intercept) is found in case of Jalna market (40604 quintals) and lowest in case of Osmanabad market (2260 quintals). The annual rate of increase in arrivals of wheat is found to be the highest in case of Aurangabad market (1909 quintals) and the lowest in case of Osmanabad (90 quintals) market during the period from 1981 to 1995. It means that the arrivals of wheat have increased at the highest annual rate in Aurangabad market and at the lowest rate in Osmanabad market during the study period. The wheat crop arrivals have enhanced on account of increase in area under cultivation in these market areas. The annual rate of increase in arrivals of wheat is found to be statistically significant in case of Aurangabad (at 5 per cent level) market. The annual rate of increase is not found statistically significant in case of Jalna, Osmanabad and Nanded markets. Among these three markets annual trend values of arrivals of wheat have shown a decreasing trend during the study period, which are: Latur, Parbhani and Beed. The highest base year value of arrivals of wheat (i.e. intercept) is found in case of Latur (41767 quintals) and the lowest in case of Parbhani (13654 quintals) market. The annual rate of decrease in arrivals of wheat is found to be the highest in case of Latur (-1899 quintals) and the lowest in case of Beed (-49 quintals) during the study period. In these market areas, the average yield per hectare of wheat has decreased and further that the farmers

Table 8
The Linear Trend Equations for Arrivals and Prices of Wheat at Regulated Markets in Marathwada During 1981-95

Markets	*Arrivals*		*Prices*	
	a	*b*	*a*	*b*
Aurangabad	172.00NS	19.09**	152.10***	22.16***
	(165.16)	(9.87)	(43.45)	(2.60)
Jalna	406.04**	7.84NS	146.39***	24.71***
	(194.78)	(11.64)	(42.54)	(2.54)
Parbhani	136.54***	-8.59***	157.57***	22.80***
	(28.18)	(1.68)	(43.18)	(2.58)
Beed	164.65**	-0.49NS	136.67***	27.62***
	(71.89)	(4.30)	(49.27)	(2.94)
Latur	417.67**	-18.99**	137.40***	26.71***
	(208.04)	(12.43)	(38.94)	(3.52)
Osmanabad	22.6NS	0.90NS	147.40***	26.26***
	(20.22)	(1.21)	(47.73)	(2.85)
Nanded	167.62*	4.63NS	141.74***	22.57***
	(107.57)	(6.43)	(41.48)	(2.48)

Note: Bracketed figures are standard errors.
a = Intercept, b = Slope,
***Significant at 1 per cent level.
**Significant at 5 per cent level.
*Significant at 10 per cent level, NS = Non-significant.

shifted emphasis from wheat to other cash crops. The annual rate of decrease in arrivals of wheat is found to be statistically significant at 1 per cent level in case of Parbhani, 5 per cent level in case of Latur and statistically insignificant in case of Beed market.

The annual trend values of prices of wheat have shown an increasing trend in all the markets. The highest base year value of price (i.e. intercept) is found in case of Parbhani (Rs. 157.57) and the lowest in case of Beed (Rs. 136.67) market. The annual rate of increase in the price of wheat is found to be the highest in case of Beed (Rs. 27.62) market and the lowest for Aurangabad (Rs. 22.16) market during the period under consideration. Further, the annual rate of increase in wheat price is statistically significant at 1 per cent level in all the markets.

Rabi Jowar: The annual trend values of arrivals and prices of rabi jowar for regulated markets are given in table 9. It is observed from this table that annual trend values of arrivals of rabi jowar have shown increasing trend over the period of study in four regulated markets. The markets where the trend values of arrivals increased include Jalna, Osmanabad, Beed and Nanded. Among these markets the highest base year value of arrivals (i.e. intercept) is found in case of Jalna (68045 quintals) and the lowest in case of Osmanabad (3716 quintals). The annual rate of increase in arrivals of this commodity is found to be the highest in case of Jalna (3432 quintals) market and the lowest in case of Osmanabad (243 quintals) market. It means that the arrivals of rabi jowar have increased at the highest annual rate in Jalna market and at the lowest rate in Osmanabad market during the study period. These markets showed increased arrivals of the crop which may be attributed to increase in production and the farmers expectations for better prices. The annual rate of increase in arrivals of rabi jowar is found to be statistically significant in case of Osmanabad, Nanded (at 5 per cent level) and Jalna (at 10 per cent level) markets. The annual rate of increase in arrivals is not found statistically significant in case of Beed market. Among the markets annual trend values of arrivals of rabi jowar have shown a decreasing trend during the study period. These markets are: Aurangabad, Latur and Parbhani. Among these markets the highest base year value of arrivals of rabi jowar (i.e. intercept) is found in case of Latur (27494 quintals) market and the lowest in case of Aurangabad (-278.35 quintals) market. The annual rate of decrease in arrivals of rabi jowar is found to be the highest in case of Parbhani (-1162 quintals) market and the lowest in case of Aurangabad (-304 quintals) market during the study period. The decreasing trend is observed due to the reduced production of the crop and poor weather conditions. The annual rate of decrease in arrivals of rabi jowar is found statistically significant at 1 per cent level in case of Parbhani market, and statistically insignificant in case of Aurangabad and Latur markets.

Table 9

Trend Values of Arrivals and Prices of Rabi Jowar at Regulated Markets in Marathwada During 1981-95

(Arrivals in '00 Quintals)
(Price in Rs. per Quintal)

Years	*Markets*													
	Aurangabad		*Parabhani*		*Jalna*		*Nanded*		*Latur*		*Beed*		*Osmanabad*	
	Trends values of arrivals	*Trends values of prices*	*Trends values of arrivals*	*Trends values of prices*	*Trends values of arrivals*	*Trends values of prices*	*Trends values of arrivals*	*Trends values of prices*	*Trends values of arrivals*	*Trends values of prices*	*Trends values of arrivals*	*Trends values of prices*	*Trends values of arrivals*	*Trends values of prices*
1981	275.31	131.87	165.06	143.29	714.77	124.71	138.26	104.64	265.32	127.43	399.59	115.32	39.59	137.08
1982	272.27	157.74	153.44	166.95	749.09	150.99	156.40	119.92	255.65	156.49	408.43	142.87	42.02	162.74
1983	269.23	183.61	141.82	190.61	783.41	177.27	174.54	135.20	245.98	185.55	417.27	170.42	44.45	188.40
1984	266.99	209.48	130.20	214.27	817.73	203.55	192.68	150.48	236.11	214.61	426.11	197.97	46.88	214.06
1985	263.15	235.35	118.58	237.93	852.05	229.83	210.82	165.76	226.64	243.67	434.95	225.52	49.31	239.72
1986	260.11	261.22	106.96	261.59	886.37	256.11	228.96	181.04	216.97	272.73	443.79	253.07	51.74	265.38
1987	257.07	287.09	95.34	285.25	920.69	282.39	247.10	196.32	207.30	301.79	452.63	280.62	54.17	291.04
1988	254.03	312.96	83.72	308.91	955.01	308.67	265.24	211.60	197.63	330.85	461.47	308.17	56.60	316.70
1989	250.99	338.83	72.10	332.57	989.33	334.95	283.38	226.88	187.96	359.91	470.31	335.72	59.03	342.36
1990	247.95	364.70	60.48	356.23	1023.65	361.23	301.52	242.16	178.29	388.97	479.15	363.27	61.46	368.02
1991	244.91	390.57	48.86	379.89	1057.97	387.51	319.66	257.44	168.62	418.03	487.99	390.82	63.89	393.68
1992	241.87	416.44	37.24	403.55	1092.29	413.79	337.80	272.72	158.95	447.09	496.83	418.37	66.32	419.34
1993	238.83	442.31	25.62	427.21	1126.61	440.07	355.94	288.00	149.28	476.15	505.67	445.92	68.75	445.00
1994	235.79	468.18	14.00	450.87	1160.93	466.35	374.08	303.28	139.61	405.21	514.51	473.47	71.18	470.66
1995	232.75	494.05	2.38	474.53	1195.25	492.63	392.22	318.56	129.94	534.27	523.35	501.02	73.61	496.32

Note: Trend values of arrivals and prices are computed by employing the least-squares method.
Source: As indicated in table 7.

The annual trend values of prices of rabi jowar have shown an increasing trend in all the markets during the study period. The highest base year value of price of rabi jowar (i.e. intercept) is found in case of Parbhani (Rs. 119.63) and the lowest in case of Beed (Rs. 87.77) market. The annual rate of increase in the prices of rabi jowar is found to be the highest in case of Latur (Rs. 29.06) market and the lowest for Nanded (Rs. 15.28) market. Further, the annual rate of increase in rabi jowar prices is statistically significant at 1 per cent level in all the markets.

Table 10
The Linear Trend Equations for Arrivals and Prices of Rabi Jowar at Regulated Markets in Marathwada During 1981-95

Markets	*Arrivals*		*Prices*	
	a	*b*	*a*	*b*
Aurangabad	-278.35***	-3.04NS	106.00*	25.87***
	(94.25)	(5.63)	(74.09)	(4.43)
Jalna	680.45**	34.32*	98.43*	26.28***
	(332.56)	(19.87)	(72.63)	(4.34)
Parbhani	176.68***	-11.62***	119.63**	23.66***
	(36.80)	(2.20)	(58.77)	(3.51)
Beed	390.75**	8.84NS	87.77NS	27.55***
	(153.93)	(9.20)	(80.84)	(4.83)
Latur	274.94**	-9.67NS	98.37NS	29.06***
	(129.23)	(7.72)	(76.88)	(4.59)
Osmanabad	37.16*	2.43**	111.42*	25.66***
	(22.52)	(1.35)	(74.58)	(4.46)
Nanded	120.12NS	18.14**	89.36*	15.28***
	(151.39)	(9.05)	(62.65)	(3.74)

Note: Bracketed figures are standard errors.
a = Intercept, b = Slope,
***Significant at 1 per cent level, ** Significant at 5 per cent level,
*Significant at 10 per cent level, NS = Non-significant.

Kharif Jowar: The annual trend values of arrivals and prices of kharif jowar for regulated markets are given in table 11. It is observed from this table that annual trend values of arrivals of kharif jowar have shown increasing trend over the period of this study only in Latur regulated market . The markets where annual trend values of arrivals showed increasing trend, the base year value of arrivals (i.e. intercept) is found in case of Latur (178629 quintals) market and the annual rate of increase is found to be the highest in case of Latur (1095 quintal) market again. The annual rate of increase in arrivals of kharif jowar is found to be statistically insignificant in case of Latur market only.

Among the five markets annual trend values of kharif jowar arrivals have shown a decreasing trend. These markets are : Aurangabad, Jalna, Osmanabad, Parbhani and Beed. The highest base year value of arrival of kharif jowar (i.e. intercept) is found in case of Jalna (85545 quintals)

Table 11

Trend Values of Arrivals and Prices of Kharif Jowar at Regulated Markets in Marathwada During 1981-95

(Arrivals in '00 Quintals)
(Price in Rs. per Quintal)

Years	*Markets*											
	Aurangabad		*Osmanabad*		*Jalna*		*Parabhani*		*Latur*		*Beed*	
	Trends values of arrivals	*Trends values of prices*	*Trends values of arrivals*	*Trends values of prices*	*Trends values of arrivals*	*Trends values of prices*	*Trends values of arrivals*	*Trends values of prices*	*Trends values of arrivals*	*Trends values of prices*	*Trends values of arrivals*	*Trends values of prices*
1981	387.23	90.04	125.22	88.53	827.20	88.56	152.12	74.89	1797.24	79.84	133.02	88.92
1982	372.44	104.84	121.32	104.05	798.95	104.33	143.79	89.09	1808.19	96.92	124.34	105.69
1983	357.65	119.64	117.42	119.57	770.70	120.10	135.46	103.29	1819.14	114.00	115.66	122.46
1984	342.86	134.44	113.52	135.09	742.45	135.87	127.13	117.49	1830.09	131.08	106.98	139.23
1985	328.07	149.24	109.62	150.61	714.20	151.64	118.80	131.69	1841.04	148.16	98.30	156.00
1986	313.28	164.04	105.72	166.13	685.95	167.41	110.47	145.89	1851.99	165.24	89.62	172.77
1987	298.49	178.84	101.82	181.65	657.70	183.18	102.14	160.09	1862.94	182.32	80.94	189.54
1988	283.70	193.64	97.92	197.17	629.45	198.95	93.81	174.29	1873.89	199.40	72.26	206.31
1989	268.91	208.44	94.02	212.69	601.20	214.72	85.48	188.49	1884.84	216.48	63.58	223.08
1990	254.12	223.24	90.12	228.21	572.95	230.49	77.15	202.69	1895.79	233.56	54.90	239.85
1991	239.33	238.04	86.22	243.73	544.70	246.26	68.82	216.89	1906.74	250.64	46.22	256.62
1992	224.54	252.84	82.32	259.25	516.45	262.03	60.49	231.09	1917.69	267.72	37.54	273.39
1993	209.75	267.64	78.42	274.77	488.20	277.80	52.16	245.29	1928.64	284.80	28.86	290.16
1994	194.96	282.44	74.52	290.29	459.95	293.57	43.83	259.49	1939.59	301.88	20.18	306.93
1995	180.17	297.24	70.62	305.81	431.70	309.34	35.50	273.69	1950.54	318.96	11.50	323.70

Note: Trend values of arrivals and prices are computed by employing the least-squares method.
Source: As indicated in table 7.

and the lowest in case of Osmanabad (12912 quintals) market. Further, the annual rate of decrease in arrivals of kharif jowar is found to be the highest in case of Jalna (-2825 quintals) and the lowest in case of Osmanabad (-390 quintals) market. The decreasing trend found in kharif jowar arrivals in most of the markets is due to shift in the area under cultivation from this crop to other crops. The annual rate of decrease in arrivals of kharif jowar is found to be statistically significant at 1 per cent level in case of Beed, at 5 per cent level in case of Parbhani and Aurangabad and at 10 per cent level in case of Osmanabad and Jalna markets.

Table 12
The Linear Trend Equations for Arrivals and Prices of Kharif Jowar at Regulated Markets in Marathwada During 1981-95

Markets	*Arrivals*		*Prices*	
	a	*b*	*a*	*b*
Aurangabad	402.02***	-14.79**	75.24NS	14.80***
	(105.40)	(6.30)	(78.63)	(4.70)
Jalna	855.45***	-28.25*	72.79NS	15.77***
	(306.64)	(18.33)	(69.76)	(4.17)
Parbhani	160.45***	-8.33**	60.69NS	14.20***
	(56.42)	(3.37)	(63.13)	(3.77)
Beed	141.70***	-8.68***	72.15NS	16.77***
	(37.59)	(2.25)	(70.13)	(4.19)
Latur	1786.29***	10.95NS	62.76NS	17.08***
	(681.05)	(40.70)	(65.76)	(3.93)
Osmanabad	129.12***	-3.90*	73.01NS	15.52***
	(42.32)	(2.53)	(63.28)	(3.78)

Note: Bracketed figures are standard errors.
a = Intercept, b = Slope,
***Significant at 1 per cent level, ** Significant at 5 per cent level,
*Significant at 10 per cent level, NS = Non-significant.

The annual trend values of prices of kharif jowar have shown an increasing trend in all the markets. The highest base year value of price of kharif jowar (i.e. intercept) is found in case of Osmanabad (Rs. 73.01) and the lowest in case of Parbhani (Rs. 60.69) market. The annual rate of increase in the price of kharif jowar is found to be the highest in case of Latur (Rs. 17.08) market and the lowest for Parbhani (Rs. 14.20) market during the period under consideration. Further, the annual rate of increase in kharif jowar price is statistically significant at 1 per cent level in all the markets.

Tur: The annual trend values of arrivals and prices of tur for selected regulated markets are given in table 13. It is observed from this table that annual trend values of arrivals of tur have shown a rising trend over the period of this study in four regulated markets viz., Aurangabad, Jalna, Osmanabad and Beed. Among the markets where annual trend values of arrivals of tur showed rising trend the highest base year value

Table 13

Trend Values of Arrivals and Prices of Tur at Regulated Markets in Marathwada During 1981-1995

(Arrivals in '00 Quintals)
(Price in Rs. per Quintal)

Years	Markets													
	Aurangabad		Parabhani		Jalna		Nanded		Latur		Beed		Osmanabad	
	Trends values of arrivals	Trends values of prices	Trends values of arrivals	Trends values of prices	Trends values of arrivals	Trends values of prices	Trends values of arrivals	Trends values of prices	Trends values of arrivals	Trends values of prices	Trends values of arrivals	Trends values of prices	Trends values of arrivals	Trends values of prices
1981	15.00	176.68	295.69	191.11	605.70	202.65	227.50	220.25	1595.95	208.72	215.10	189.07	62.46	230.25
1982	53.61	248.83	292.89	262.25	632.72	277.40	223.57	290.33	1586.37	286.25	223.41	265.19	66.36	299.20
1983	92.22	320.98	290.09	333.39	659.74	352.15	219.64	360.41	1576.98	363.78	231.72	341.31	70.26	368.15
1984	130.83	393.13	287.29	404.53	686.76	426.90	215.71	430.49	1567.59	441.31	240.03	417.43	74.16	437.10
1985	169.44	465.28	284.49	475.67	713.78	501.65	211.78	500.57	1558.20	518.84	248.34	493.55	78.06	506.05
1986	208.05	537.43	281.69	546.81	740.80	576.40	207.85	570.65	1548.81	596.37	256.65	569.67	81.96	575.00
1987	246.66	609.58	278.89	617.95	767.82	651.15	203.92	640.73	1539.42	673.90	264.96	645.79	85.86	643.95
1988	285.27	681.73	276.09	689.09	794.84	725.90	199.99	710.81	1530.03	751.43	273.27	721.91	89.76	712.90
1989	323.88	753.88	273.29	760.23	821.86	800.65	196.06	780.89	1520.64	828.96	281.58	798.03	93.66	781.85
1990	362.49	826.03	270.49	831.37	848.88	875.40	192.13	850.97	1511.25	906.49	289.89	874.15	97.56	850.80
1991	401.10	898.18	267.69	902.51	875.90	950.15	188.20	921.05	1501.86	984.02	298.20	950.27	101.46	919.75
1992	439.71	970.33	264.89	973.65	902.92	1024.90	184.27	991.13	1492.47	1061.55	306.51	1026.39	105.36	988.70
1993	478.32	1042.48	262.09	1044.79	929.94	1099.65	180.34	1061.21	1483.08	1139.08	314.82	1102.51	109.26	1057.65
1994	516.93	1114.63	259.29	1115.93	956.96	1174.40	176.41	1131.29	1473.69	1216.61	323.13	1178.63	113.16	1126.60
1995	555.54	1564.67	256.49	1187.07	983.98	1249.15	172.48	1201.37	1464.30	1294.14	331.44	1254.75	117.06	1195.55

Note: Trend values of arrivals and prices are computed by employing the least-squares method.
Source: As indicated in table 7.

of arrivals of this commodity (i.e. intercept) is found in case of Jalna (57868 quintals) market and the lowest in case of Aurangabad (-2361 quintals) market. The annual rate of increase is found to be the highest in case of Aurangabad (3861 quintals) market and the lowest in case of Osmanabad (390 quintals) market. It means that the arrivals of tur have increased at the highest annual rate in Aurangabad market and at the lowest rate in Osmanabad market during the study period. Due to good production of the crop and attractive prices in the market, has resulted in the increased arrivals. The annual rate of increase in arrivals of tur is found to be statistically significant at 1 per cent level in case of Aurangabad market, at 5 per cent level in case of Osmanabad market, and not found statistically significant in case of Jalna and Beed markets.

Annual trend values of arrivals of tur have shown a decreasing trend in Parbhani, Nanded and Latur markets. The highest base year value of arrivals of tur (i.e. intercept) is found in case of Latur (160515 quintals) market and the lowest in case of Nanded (23143 quintals) market. The annual rate of decrease in arrivals of tur is found to be the highest In case of Latur (-933 quintals) market and the lowest in case of Parbhani (-280 quintals) market. The annual rate of decrease in arrivals of tur is found statistically insignificant in case of Latur, Parbhani and Nanded markets.

The annual trend values of prices of tur have shown an increasing trend in all the markets during the study period. The highest base year value of price for this commodity (i.e. intercept) is found in case of Osmanabad (Rs. 161.30) and the lowest in case of Aurangabad (Rs. 104.53) market. The annual rate of increase in the price of tur is found to be the highest in case of Latur (Rs. 77.53) market and the lowest for Osmanabad (Rs. 68.95) market. Further, the annual rate of increase in the prices is statistically significant at 1 per cent level in all the markets (see table 14).

Safflower: The annual trend value of arrivals and prices of safflower are shown in table 15. From this table we can see that the annual trend values of the arrivals of safflower have shown a decreasing trend in all the six markets. The annual arrivals of safflower have decreased in all the markets over the period under consideration. This particular result may be due to low production of this commodity in the areas of the regulated markets in Marathwada. The base year value of the arrival of safflower is found to be the highest for Jalna (i.e. 60461 quintals) market and the lowest for Osmanabad (i.e. 946 quintals) market. Though the trend values of the arrivals of safflower have decreased in all the markets, the annual rate of decrease in the arrivals of safflower is highest in case of Jalna (i.e. -3107 quintals) market and the lowest for Osmanabad (i.e. -47 quintals) market. The highest rate of decrease in arrivals of safflower at Jalna market may be due to fast decrease in production of this commodity in the surrounding area of this market. Due to insufficient monsoons, the acreage under cultivation gets reduced and it finally reduces the production which adversely affects the market arrivals. The rate of annual decrease in the arrivals of safflower is found statistically non significant in Latur market, 1 per cent level in case of Aurangabad, Jalna, Parbhani and Beed markets and at 5 per cent level in case of Osmanabad market.

Table 14
The Linear Trend Equations for Arrivals and Prices of Tur at Regulated Markets in Marathwada During 1981-95

Markets	*Arrivals*		*Prices*	
	a	*b*	*a*	*b*
Aurangabad	-23.61^{NS}	38.61***	104.53^{NS}	72.15***
	(101.41)	(6.06)	(142.47)	(8.51)
Jalna	578.68*	27.02^{NS}	127.90^{NS}	74.75***
	(336.47)	(20.11)	(141.39)	(8.45)
Parbhani	298.49***	-2.80^{NS}	119.97^{NS}	71.14***
	(100.99)	(6.04)	(118.41)	(7.06)
Beed	206.79*	8.31^{NS}	112.95^{NS}	76.12***
	(139.52)	(8.34)	(156.57)	(9.36)
Osmanabad	58.56**	3.90**	161.30^{NS}	68.95***
	(30.03)	(1.79)	(126.91)	(7.58)
Latur	1605.15***	-9.33^{NS}	131.19^{NS}	77.53***
	(553.01)	(33.05)	(157.97)	(9.44)
Nanded	231.43**	-3.93^{NS}	150.17^{NS}	70.08***
	(94.59)	(5.65)	(133.86)	(8.00)

Note: Bracketed figures are standard errors.
a = Intercept, b = Slope,
***Significant at 1 per cent level, **Significant at 5 per cent level,
*Significant at 10 per cent level, NS = Non-significant.

The annual trend values of prices for safflower have shown a rising trend in all the six regulated markets. The base year price of safflower is as low as Rs. 232.33 per quintal in Latur market and as high as Rs. 335.33 per quintal in Parbhani market. The rate of increase in prices of safflower differed from market to market. However, it was found to be the highest in case of Jalna (Rs. 51.97 per quintal) market and the lowest (Rs. 31.17 per quintal) in case of Parbhani market. Further, it is important to note that this rate of increase in price of safflower is found highly significant at 1 per cent level in all the regulated markets under consideration during the study period.

Groundnut: The annual trend values of arrivals and prices of groundnut are shown in table 17. From this table it is clear that the arrivals of groundnut have shown a rising trend in three markets viz.; Parbhani, Nanded and Beed and a decreasing trend in three markets viz.; Jalna, Latur and Osmanabad. Among the markets where the

Table 15
Trend Values of Arrivals and Prices of Safflower at Regulated Markets in Marathwada During 1981-1995

(Arrivals in '00 Quintals)
(Price in Rs. per Quintal)

Years	Markets											
	Aurangabad		Parabhani		Jalna		Beed		Latur		Osmanabad	
	Trends values of arrivals	Trends values of prices	Trends values of arrivals	Trends values of prices	Trends values of arrivals	Trends values of prices	Trends values of arrivals	Trends values of prices	Trends values of arrivals	Trends values of prices	Trends values of arrivals	Trends values of prices
1981	162.70	302.71	144.14	366.50	573.54	305.67	197.40	332.81	109.89	283.15	8.99	324.21
1982	153.79	354.04	134.21	397.67	542.47	357.64	184.26	372.20	108.06	333.97	8.52	257.36
1983	144.88	405.37	124.28	428.84	511.40	409.61	171.12	411.59	106.23	384.79	8.05	390.51
1984	135.97	456.70	114.35	460.01	480.33	461.58	157.98	450.98	104.40	435.61	7.58	423.66
1985	127.06	508.03	104.42	491.18	449.26	513.55	144.84	490.37	102.57	486.43	7.11	456.81
1986	118.15	559.36	94.49	522.35	418.19	565.52	131.70	529.76	100.74	537.25	6.64	489.96
1987	109.24	610.69	84.56	553.52	387.12	617.49	118.56	569.15	98.91	588.07	6.17	523.11
1988	100.33	662.02	74.63	584.69	356.05	669.46	105.42	608.54	97.08	638.89	5.70	556.26
1989	91.42	713.35	64.70	615.86	324.98	721.43	92.28	647.93	95.25	689,71	5.23	589.41
1990	82.51	764.68	54.77	647.03	293.91	773.40	79.14	687.32	93.42	740.53	4.76	622.56
1991	73.60	816.01	44.84	678.20	262.84	825.37	66.00	726.71	91.59	791.35	4.29	655.71
1992	64.69	867.34	34.91	709.37	231.77	877.34	52.86	766.10	89.76	842.17	3.82	688.46
1993	55.78	918.67	24.98	740.54	200.70	929.31	39.72	805.49	87.93	892.99	3.35	722.01
1994	46.87	970.00	15.05	771.71	169.63	981.28	26.58	844.88	86.10	943.81	2.88	755.16
1995	37.96	1021.33	5.12	802.88	138.56	1033.25	13.44	884.27	84.27	994.63	2.41	788.81

Note: Trend values of arrivals and prices are computed by employing the least-squares method.
Source: As indicated in table 7.

Table 16
The Linear Trend Equations for Arrivals and Prices of Safflower at Regulated Markets in Marathwada During 1981-95

Markets	*Arrivals*		*Prices*	
	a	*b*	*a*	*b*
Aurangabad	171.61*** (43.77)	-8.91*** (2.62)	251.33** (102.12)	51.33*** (6.10)
Jalna	604.61*** (175.88)	-31.07*** (10.51)	253.70** (99.08)	51.97*** (5.92)
Parbhani	154.07*** (40.17)	-9.93*** (2.40)	335.33*** (124.50)	31.17*** (7.44)
Beed	210.54*** (47.23)	-13.14*** (2.82)	293.42*** (77.76)	39.39*** (4.65)
Latur	111.72*** (31.88)	-1.83NS (2.22)	232.33** (99.22)	50.82*** (5.93)
Osmanabad	9.46** (4.38)	-0.47** (0.26)	291.06*** (99.67)	33.15*** (5.96)

Note: Bracketed figures are standard errors.
a = Intercept, b = Slope,
***Significant at 1 per cent level, **Significant at 5 per cent level,
*Significant at 10 per cent level, NS = Non-significant.

arrivals have shown rising trend, the highest annual rate of increase in arrivals of this commodity is found in case of Beed (16158 quintals) and the lowest in case of Parbhani (.345 quintals) market. The assured irrigation during summer season has attracted farmers to take this crop on a larger scale, and the prices of groundnut are also attractive. The yield of the crop is further satisfactory which ultimately results in enhanced production and market arrivals. The annual rate of increase in arrivals of this commodity is found statistically significant at 5 per cent level in case of Parbhani and Nanded markets while in Beed market it is found to be insignificant.

The arrivals of groundnut have shown the highest rate of annual decrease for Latur (-1788 quintals) and the lowest for Osmanabad (-77 quintals). The annual rate of decrease in arrivals for Jalna and Latur markets are statistically significant at 5 per cent level but for Osmanabad it is statistically insignificant.

The prices of groundnut have shown a rising trend during the study period in all the six markets. However, the rate of increase in the price of groundnut differed from market to market. It is found to be the highest in case of Nanded market (Rs. 68.17 per quintal) and the lowest for Parbhani (Rs. 50.17 per quintal) market. It is important to note here that the rate of increase in prices of groundnut is found to be highly significant (at 1 per cent level) in all the markets.

Prices for all the agricultural commodities from all the markets have shown an increasing trend. This may be attributed to the factors such

Table 17

Trend Values Of Arrivals And Prices Of Groundnut At Regulated Markets In Marathwada During 1981-1995

(Arrival in '00 Quintals)
(Price in Rs. per Quintal)

Years	*Markets*											
	Jalna		*Latur*		*Parabhani*		*Osmanabad*		*Beed*		*Nanded*	
	Trends values of arrivals	*Trends values of prices*	*Trends values of arrivals*	*Trends values of prices*	*Trends values of arrivals*	*Trends values of prices*	*Trends values of arrivals*	*Trends values of prices*	*Trends values of arrivals*	*Trends values of prices*	*Trends values of arrivals*	*Trends values of prices*
1981	86.24	206.07	611.15	318.62	6.00	283.20	33.72	321.13	164.34	285.37	115.73	251.48
1982	80.96	265.03	593.27	383.32	15,45	333.37	32.95	378.63	167.10	351.58	139.47	319.65
1983	75.68	323.99	575.39	448.02	24.90	383.54	32.18	436.13	169.86	417.79	163.21	387.82
1984	70.40	382.95	557.51	512.72	34.35	433.71	31.41	493.63	172.62	484.00	186.95	455.99
1985	65.12	441.91	539.63	577.42	43.80	483,88	30.64	551.13	175.38	550.21	210.69	524.16
1986	59.84	500.87	521.75	642.12	53.25	534.05	29.87	608.63	178.14	616.42	234.43	592.33
1987	54.56	559.83	503.87	706.82	62.70	584.22	29.10	666.13	180.90	682.63	258.17	660.50
1988	49.28	618.79	485.99	771.52	72.15	634.39	28.33	723.63	183.66	748.84	281.91	728.67
1989	44.00	677.75	468.11	836.22	81.60	684.56	27.56	781.13	186.42	815.05	305.65	796.84
1990	38.72	736.71	450.23	900.92	91.05	734.73	26.79	838.63	189.18	881.26	329.39	865.01
1991	33.44	795.67	432.35	965.62	100.50	784.90	26.02	896.13	191.94	947.47	353.13	933.18
1992	28.16	854.63	414.47	1030.32	109.95	835.07	25.25	953.63	194.70	1013.68	376.87	1001.35
1993	22.88	913.59	396.59	1095.02	119.40	885.24	24,48	1011.13	197.46	1079.89	400.61	1069.52
1994	17.60	972.55	378.71	1159.72	128.85	935.41	23.71	1068.63	200.22	1146.10	424.35	1137.69
1995	12,32	1031.51	360.83	1224.42	138.30	985.58	22.94	1126.13	202.98	1212.31	448.09	1205.86

Note: Trend values of arrivals and prices are computed by employing the least-squares method.
Source: As indicated in table 7.

as population growth, changes in the weather conditions, increase in per capita income of the consumers and rising cost of inputs in farming.

Table 18
The Linear Trend Equations for Arrivals and Prices of Groundut at Regulated Markets in Marathwada During 81-95

Markets	*Arrivals*		*Prices*	
	a	*b*	*a*	*b*
Jalna	91.52*** (33.75)	-5.28** (2.02)	147.11NS (114.23)	58.96*** (6.83)
Parbhani	-3.45NS (65.93)	9.45** (3.94)	233.03* (163.95)	50.17*** (9.80)
Beed	161.58** (84.62)	2.76NS (5.06)	219.16** (109.27)	66.21*** (6.53)
Latur	629.03*** (134.08)	-17.88** (8.01)	253.92** (102.52)	64.70*** (6.13)
Osmanabad	34.49** (14.91)	-0.77NS (0.89)	263.63*** (94.29)	57.50*** (5.64)
Nanded	91.99NS (215.47)	23.74** (12.88)	183.31NS (139.04)	68.17*** (8.31)

Note: Bracketed figures are standard errors.
a = Intercept, b = Slope,
***Significant at 1 per cent level, **Significant at 5 per cent level,
*Significant at 10 per cent level, NS = Non-significant.

CONCLUSIONS

1. This study has indicated that the seasonal variation in arrivals of wheat were higher in Jalna and Nanded markets and lower in Osmanabad and Aurangabad markets. Kharif jowar arrivals have shown the highest degree of seasonal variation in Latur and Parbhani markets and the lowest variation in Beed and Latur markets, Rabi jowar shown a higher degree of variation in Jalna market and lower degree of variation in Parbhani market.
2. Arrivals of tur have been subject to larger seasonal variation in Latur and Aurangabad markets and smaller variation in Osmanabad and Nanded markets and in case of safflower seasonal variation in arrivals was the highest in case of Jalna and Parbhani markets and the lowest in Osmanabad and Jalna markets. In case of the groundnut seasonal variation of arrivals was the highest in Nanded and lowest in Osmanabad and Latur markets. But the variation in its seasonal prices was the highest in Jalna and lowest in Beed market.
3. Arrivals of wheat have shown a rising trend in seven markets and a falling trend in another three markets. The annual rate of increase or decrease in arrivals is also found to be different for various

markets. The highest and the lowest annual rates of increase in arrivals of wheat were found for Aurangabad and Osmanabad markets respectively whereas the highest and lowest annual rates of decrease in arrivals were found for Latur and Beed markets respectively. Rabi jowar arrivals have shown a decreasing trend in three markets and increasing trend in only four markets. The highest annual rate of decrease in arrivals of rabi jowar was found for Latur and the lowest for Aurangabad market. The decreasing trend is observed due to the reduced production of the crop and poor weather conditions.

4. In this study it is found that the arrivals of Kharif jowar have shown a decreasing trend in all the markets except Latur. The annual rate of decrease is found to be the highest for Jalna and lowest for Osmanabad market. The decreasing trend found in kharif jowar arrivals in most of the markets is due to shift in the area under cultivation from this crop to other crops. The arrivals have recorded the highest annual rate of increase in Aurangabad and lowest rate in Osmanabad market.
5. This study has indicated that the arrivals and prices of safflower have shown opposite trends in all the markets. Its arrivals have shown a decreasing trend but its prices have shown an increasing trend during the study period. The highest annual rate of decrease in arrivals is found for Jalna and lowest for Osmanabad. Groundnut arrivals have shown a rising trend in three markets and decreasing trend in three markets. Groundnut arrivals have exhibited the highest annual rate of increase in Beed and the lowest rate of increase in Parbhani market.
6. Prices for all the crops in all the markets have shown an increasing trend. This may be attributed to the factors such as population growth, changes in the weather conditions, increase in per capita income of the consumers and rising cost of inputs in farming.

Thus the study suggests the need for providing storage, transporation, mordernisation of market committies, grading, loan facilities, role of agricultural produce market committies in price fixation, agriculturalisation of industry for minimizing the seasonal varation in the arrivals and prices of agricultural commodities. It will definitely help for stabilizing the incomes of the growers.

REFERENCES

Acharya, S.S. and Agrawal, N.L.(1994), Agricultural Prices-Analysis and Policy, Oxford and IBH Publishing Co. Pvt. Ltd. New Delhi.

Acharya, S.S. and Agrwal, N.L. (1996), Agricultural Marketing in India, Oxford and IBH Publishing Co. Pvt. Ltd., New Delhi.

Brahma Prakash and Sushila Srivastava (1993), Analysis of Market Arrivals and Prices of gram in Uttar Pradesh, Indian Journal of Agricultural Economics, Vol. 48, No. 3.

Indrasena Reddy (1995), Price Behaviour of Agricultural Commodities, Indian Journal of Marketing. Vol. 23, No. 6-7.

Babar, V.S. and Lohar, N.S. (1994), Trends in Arrivals and Prices of Jaggery in Sangli Regulated market, Indian Journal of Agricultural Marketing, Vol. 8, No. 1.

Puran Chand, Sharma B.M. and Chhatan Singh (1991), Trends in Agricultural Input Output Prices in India, Southern Economist, Vol. 29, No. 20.

Agrawal N.L. (1986), Foodgrains Prices Analysis–An Interpretative Review of Studies in India.

B.D. KHANDARE

DEVELOPMENT OF CO-OPERATIVE SUGAR INDUSTRY IN MAHARASHTRA

9

The Sugar industry is the second largest agro-based industry in India to textile. Though the New Guinea and its neighbouring islands in the South Pacific are now widely believed to be the natural home of sugarcane.[1] Sugar was first produced in India as early as 400 B.C.when the process had become generally known.[2] Noel Deer mention that the earliest datable mention of sugarcane is 325 B.C. when it is recorded by Nearchus, the Greek recorder in western India. In his words, "The production of sugar and syrup was known in India perhaps a few centuries before the Christianera and certainly by 400 A.D.[3]

Maharashtra state is the sugar bowl of India. No effort was made to organise the co-operative sugar factories in the old Bombay state which is a part of the Maharashtra state now till 1947. It was only in 1948, the first co-oprative sugar factory was established in Ahmednagar district at Pravaranagar in 1948 was itself a very good beginning and there after many co-operative sugar factories are developed in the state of Maharashtra. There were in all 136 sugar factories in operation during the year 2003-2004 in the state.Out of these, 124 are in cooperative sector and remaining 12 are in private sector. Over a period of last 50 years the industry has achieved a substantial quantitative progress in the country in general and in the state of Maharashtra in particular.

The climatic conditions differ from zone to zone which have affected the cultivation practices, sugarcane yield and sucrose recovery too. Naturally it effects the working results of the sugar factories in these three zones differently. The growth of sugar indusry in Marathwada region is late as compared to its counter i.e. south and cenetral zones. The sugarcane supply position varies from zone to zone, which has affected the utilization of plant capacity from zone to zone also. No doubt in south and cenrtral zone various schemes of irrigation are in vogue. But the Marathwada region no such schemes are implemented except few pockets. The North-East zone of Maharashtra consists of three sub regions viz.

1. Khandesh Region : 11 Sugar factories.
2. Vidarbha Region : 14 Sugar factories.
3. Marathwada Region : 36 Sugar factories.

The Marathwada region covers eight districts namely Aurangabad, Jalna, Parbhani, Hingoli, Nanded Beed, Osmanabad and Latur having 36 Sugar factories.

The objectives set for the study are (i) to take a review of the price policies of the central and state government towards the sugar industry with paricular reference to cooperative sugar factories, and (ii) to study the trends in area, production, yield, sugarcane crushed, production of sugar, capacity utilization, prices paid by cooperative sugar factories and recovery of sugar industry in Maharashtra.

RESEARCH METHODOLOGY

Secondary data about area, production and yield of sugarcane in factory areas, sugarcane crushed, sugar production, sugarcane prices paid by the factories, capacity utilization, sugar recovery etc. were collected from the annual reports of the sugar factories for the period 1981-82 to 2003-2004. More-over, some data were also collected from the office of the Joint Director of Sugar, Government of Maharashtra, Pune, Bhartiya Sugar Directry, Pune and Reports of the Maharashtra Rajya Sahakari Karkhana Sangh Ltd., Mumbai. In addition to this, some required secondary data were obtained from various periodicals and reports such as Districtwise General Statistical Information of Maharashtra state issued by the Directorate of Agriculture, Godwa (monthly issue), Working Results of sugar factories in Maharhashtra, Cooperative sugar, India sugar and Handbook of Basic Stastitics of Maharashtra state.

GROWTH OF SUGAR INDUSTRY IN INDIA AND MAHARASHTRA

Sugarcane and its main products are known in India since ancient times. The earliest reference of sugar appears in Atharva Veda, the sacred book of Hindus composed between 5000 and 1000 B.C. Efforts to establish a modern sugar industry however go back only to the begining of the twentieth century. India lies entirely to the north of equator. Hence, sugarcane is grown in nearly all parts of India.[4]

Sugarcane crops are grown in several rainfall,temprature and soil regions. However,sugarcane growing area could be grouped into subtropical and tropical belt.[5] The states of Maharashtra, Gujarat, Andhra Pradesh, Tamilnadu and Kerla comprise the 'tropical region' while the states of Punjab, Uttar Pradesh, Bihar and West Bengal comprise the sub-tropical region.

India is the largest producer of sugarcane in the world. There has been considerable importance to this commercial crop during the last two decades.[6] In the year 2003-04 at all India level, the area under sugarcane was 39.95 lakh hectares and production 2362 lakh tonnes with an average yield of 59.1 tonnes.

Maharashtra is one of the leading sugarcane growing states in India. It is a perennial irrigated crop grown in all the zones of the state. Maharashtra state occupied a well recognised place of pride on the sugarcane map of the country. The area under sugarcane in the Maharashtra state is 5.26 lakh hectares and have total production of 269.82 lakh tonnes with an average yield of 51.30 tonnes per hectare in the year 2003-04. The area under sugarcane in Maharashtra is concentrated in low rainfall regions in the eastern part of the Sahyadri mountains Pune division i.e.Ahmednagar, Pune and Solapur districts; and Kolhapur division i.e., Satara, Sangli and Kolhapur districts; Nasik district are the major sugarcane growing districts of the state with 71.96 per cent of the state area. Some districts of Marathwada i.e.Aurangabad and Osmanabad have moderate area under sugarcane i.e. 9.60 per cent of the state area.[7]

Table 1 explains the growth of sugar industry in India from 1981-82 to 2003-04. Thereafter, sugar production declined to 106.09 tonnes in 1992-93 and further to 98.33 in 1993-94. But in the subsequent year it jumped to 185.37 tonnes and further to 210.46 tonnes in 2002-03. In 2003-04, the sugar produced by the sugar factories came down to 135.46 tonnes. On an average, the sugar production was 119.96 lakh tonnes per year. The annual average growth rate of sugar production was 0.52 per cent.

During the study period, on an average, the recovery percentage of sugar in India was about 10 per cent. The average recovery was 10.03 Per cent. The annual average growth rate of sugar recovery was 0.08 per cent. During 1981-82, the area under sugarcane was 31.93 lakh hectares, which went up to 41.47 lakh hectares in 1996-97. During the sixteen year period, on an average, the area under sugarcane was 36.75 lakh hectares per year. The area under sugarcane was the lowest in 1985-86 (i.e. 28.50 lakh hectares) due to drought condition, and it was the highest in 2001-02 (i.e. 44.30 lakh hectares) due to favourable weather conditions. The average annual growth rate of sugarcane area was 0.17 per cent.

The average sugarcane yield in India showed increasing trend during the period under study. During 1981-82 the yield of sugarcane was 54.8 tonnes per hectare, which increased and reached up to 71.2 tonnes per hectare in 1998. On an average yield of sugarcane was 64.21 tonnes per hectare during the study period. The annual average growth rate of yield of sugarcane was 0.09 per cent. The sugarcane production in India

Table 1
Growth of Sugar Industry in India During 1981-82 to 2003-04

Year	*Area under sugarcane (Lakh hectares)*	*Sugarcane production (Lakh metric Tonnes)*	*Yield (tonnes/ hectare)*	*Sugarcane crushed (Lakh metric tonnes)*	*Sugar production (Lakh metric tonnes)*	*Average sugar recovery (per cent)*
1981-82	31.93	1864	54.8	540.10	54.6	9.66
1982-83	33.58	1895	56.4	570.15	54.8	9.95
1983-84	31.10	1741	56.0	590.24	59.17	10.02
1984-85	29.53	1703	57.7	600.90	61.48	10.22
1985-86	28.50	1706	59.9	685.66	70.16	10.23
1986-87	30.79	1861	60.4	852.02	85.02	9.98
1987-88	32.79	1967	60.0	939.36	90.10	9.70
1988-89	33.29	2030	61.0	856.47	87.53	10.22
1989-90	34.05	2226	65.4	1111.58	109.90	9.89
1990-91	36.86	2410	66.1	1223.38	124.70	9.84
1991-92	38.44	2539	63.8	1339.50	134.05	10.01
1992-93	35.72	2280	67.1	1030.08	106.09	10.30
1993-94	34.22	2296	71.1	983.38	99.33	10.00
1994-95	38.15	2712	69.9	1474.30	146.43	9.92
1995-96	41.47	2828	67.8	1747.26	164.43	9.90
1996-97	41.74	2776	66.5	1303.79	129.05	9.90
1997-98	39.40	2795	71.1	1291.35	128.52	9.95
1998-99	40.55	2887	71.2	1575.61	155.39	9.86
1999-2000	42.20	2993	70.9	1784.94	182.00	10.20
2000-01	43.16	2960	68.6	1766.60	185.11	10.48
2001-02	44.30	2984	67.4	1803.86	185.37	10.27
2002-03	43.61	2816	64.6	1943.65	210.46	10.36
2003-04	39.95	2362	59.1	1325.11	135.46	10.22
Average	36.75	2375	64.21	1188.66	119.96	10.03
Annual simple growth rate (%)	0.17	0.19	0.09	0.33	0.52	0.08

Source: Cooperative Sugar, August 2004.

showed a positive trend during the study period. During 1981-82, the sugarcane production in India was 1864 lakh tonnes, which increased up to 2993 lakh tonnes in 1999-2000. On an average, the sugarcane production was 2375 lakh tonnes per year during the study preriod.The annual average growth rate of sugarcane production was 0.19 per cent. During 1981-82, sugarcane crushed in India was 540.10 lakh metric tonnes, which increased up to 1943.65 lakh metric tonnes in 2002-03,the sugarcane crushed showed increasing trend during the study period.On an average, the sugarcane crushed was 1188.66 lakh tonnes per year during the study period. The annual average growth rate of sugarcane crushed was 0.33 per cent.

Table 2 explains the growth of sugar industry in Maharashtra during 1981-82 to 2003-04. In 1981-82 all these Sugar factories have produced 30.26 lakh metric tonnes of sugar which increased to 65.03 lakh metric tonnes in 1999-2000 and 67.05 metric tonnes in 2000-2001. On an average 39.27 metric tonnes of sugar per year has been produced by the sugar factories in Maharashtra during 1981 to 2003. The annual average growth rate of sugar production was -0.97 per cent. During the study period, the average sugar recovery in Maharashtra was 11.10 per cent, it was around 10 per cent in India. The annual average growth rate of sugar recovery was 0.01 per cent.

Table 2
Growth of Sugar Industry in Maharashtra
During 1981-82 to 2003-04

Year	*Area under sugarcane (Lakh hectares)*	*Sugarcane production (Lakh metric Tonnes)*	*Yield (tonnes / hectare)*	*Sugarcane crushed (Lakh metric tonnes)*	*Sugar production (Lakh metric tonnes)*	*Average sugar recovery (per cent)*
1981-82	2.97	288	96.9	244.88	30.26	10.70
1982-83	3.26	313	96.3	238.20	30.25	10.95
1983-84	2.94	265	90.3	158.17	19.93	11.13
1984-85	1.92	263	90.1	185.60	23.11	11.08
1985-86	2.65	237	89.4	190.21	23.88	11.20
1986-87	2.80	241	86.1	197.77	25.39	10.98
1987-88	2.92	250	85.6	245.31	27.95	10.83
1988-89	3.14	255	81.1	238.15	26.29	11.04
1989-90	3.83	340	88.8	366.29	39.21	10.71
1990-91	4.44	384	86.5	383.44	41.19	10.89
1991-92	5.18	361	79.8	376.57	42.13	11.19
1992-93	5.80	308	76.4	297.36	33.60	11.30
1993-94	5.80	278	81.1	246.85	27.46	11.13
1994-95	5.13	442	85.5	459.97	50.25	10.92
1995-96	5.80	467	80.4	514.58	53.94	10.49
1996-97	5.16	418	81.0	310.13	34.45	11.11
1997-98	4.60	382	83.0	345.77	38.47	11.13
1998-99	5.30	472	89.0	479.00	53.37	11.16
1999-2000	5.90	531	90.1	570.95	65.03	11.40
2000-01	5.95	496	83.3	676.49	67.05	11.64
2001-02	5.78	451	78.1	483.87	56.12	11.62
2002-03	5.99	370	61.80	534.74	62.21	11.68
2003-04	5.26	270	51.30	290.39	31.71	10.94
Average	4.46	351	83.13	349.33	39.27	11.10
Annual simple growth rate (%)	0.28	-0.36	-0.78	-0.81	-0.97	0.01

Sources: 1. The Report of the Committee for Recommending a Packagec for Rehabilitation of Sick Sugar Factories in Cooperative Sector published by Government of Maharashtra, Bombay, June 1999.
2. Cooperative Sugar, August, 2004.

With the increase in number of sugar factories, the area under sugarcane has also increased in Maharashtra. In 1981-82 it was 2.97

lakh hectares which showed ups and downs in the subsequent years and reached to 5.99 lakh hectares in 2002-03. During the eighties, the sugarcane acreage in the factory areas in Maharashtra was as low as 1.92 lakh hectares in 1984-85 and as high as 3.83 lakh hectares in 1989-90. During the nineties the acreage under sugarcane in factory areas of Maharashtra showed a continuously rising trend, with the exception in the year 1995-1996 and reached to a level of 5.90 lakh hectare in 1999-2000. During the study period, the annual average growth rate of area under sugarcane was 0.28 lakh hectares.

The sugarcane production in the factory areas of Maharashtra showed a mixed trend during the period under study. It was 288 lakh tonnes in 1981-82 which went up to 467 lakh tonnes in 1995-96 and further to 531 lakh tonnes in 1999-2000. The ups and downs in sugarcane production can be attributed to weather and rainfall conditions existed in different years.The annual average growth rate of sugarcane production was -0.36 per cent.

The yield of sugarcane showed a clear-cut decreasing trend till 1988-89. It was 96.9 tonnes per hectare in 1981-82 and decreased continuously to 81.8 tonnes per hectare in 1988-89. Afterwards it showed ups and downs and reached to 90.1 tonnes per hectare in 1999-2000. During the whole period under consideration, the yield of sugarcane in factory areas of Maharashtra remained at 83.13 tonnes per hectare. The annual average growth rate of sugarcane was -0.78 per cent.

SUGAR PRICES

Table 3 indicates the price of sugarcane declared by the Government of India, and price recommended by the state Government from 1981-82 to 1999-2000. This table shows that the price recommended by the state Government in 1981-82 was only Rs. 188 per metric tonne and the Price declared by Central Government was Rs. 130 per metric tonne. That means it was Rs. 18.8 and Rs.13.0 per quintal respectively during the year 1981-82. Table 3 shows that there has been continuous rising trend in Price declared by state and central governments during 1981-82 to 1999-2000. In the decade 1981-82 to 1990-91 price recommended by State Government was increased by 82.76 per cent.(from Rs. 188 per metric tonne to Rs. 343.60 per metric tonne). In 1991-92 the price fixed by State Government showed a rising trend , it was Rs. 260 per tonne in 1991-92 which increased and reached up to Rs. 560 per tonne in 1999-2000. The increase was 115 per cent in this period. On an average, the sugarcane price fixed by state Government was Rs. 335.43 per metric tonne during the study period. As well as the Price fixed by the Central Government shows continuously raising trend during the study period. During the study period the average price of sugarcane fixed by the central Government was, Rs. 288.03 per metric tonne.

Table 3
Statement Showing the Prices of Sugarcane Recommended by the State and Central Governments During 1981-82 to 1999-2000

Year	*Price recommended by state Government*	*Price declared by Central Government*
1981-82	188.00	130.00
1982-83	191.20	130.00
1983-84	196.90	135.00
1984-85	204.20	140.00
1985-86	240.70	165.00
1986-87	250.00	170.00
1987-88	267.70	185.00
1988-89	279.90	195.00
1989-90	320.90	230.00
1990-91	343.60	230.00
1991-92	260.00	260.00
1992-93	290.00	310.00
1993-94	365.00	345.00
1994-95	425.00	391.00
1995-96	450.00	425.00
1996-97	460.00	459.00
1997-98	540.00	484.50
1998-99	540.00	527.00
1999-2000	560.00	561.00
Average	335.43	288.03

Source: Various Issues of Annual Report of Maharashtra Rajya Sahakari Sakhar Karkhana Sangh Ltd., Mumbai.

In 1981-82 the Price fixed by Central Government was Rs. 130 per metric tonne which increased and reached up to Rs. 561 per metric tonne in 1999-2000, the increase was 3.31 times. The prices fixed by Central and State Government showed a positive trend during the period from 1981-82 to 1999-2000.

Table 4
Zone-wise percentage utilization of plant capacity in Maharashtra.

Year	*Marathwada*	*South zone*	*Central zone*	*Maharashtra zone*
1996-97	66.37	98.09	95.24	85.83
1997-98	71.54	95.99	95.29	87.31
1998-99	77.80	98.88	94.14	90.24
1999-00	82.88	100.46	97.25	90.86
2000-01	94.74	104.05	103.77	99.36
Average per cent	79.00	99.00	96.00	91.00

Source: Cooperative Sugar, vol. 34, September, 2002.

CAPACITY UTILIZATION

Table 4 reveals that Marathwada region has registered the lowest percentage of utilization of plant capacity as compared to south and central zones as well as at State level. South zone has recorded highest percentage of utilization of plant capacity i.e. 99 per cent, followed by Central zone i.e. 96 per cent. But the under utilization of plant capacity in Marathwada on an average was 21 per cent. If the same is compared to south and central zone it was less by 20 per cent and 17 per cent respectively.

Table 5
The statement Showing the Prices of Sugarcane Recommended by CACP and SMP Fixed by the Government of India

Year	*Price recommended by CACP (Rs. per quintal)*	*SMP fixed by the Central Government (Rs. per quintal)*
1967-68	5.68	7.37
1968-69	7.23	7.37
1969-70	7.27	7.37
1970-71	7.35	7.37
1971-72	7.37	7.37
1972-73	8.00	8.00
1973-74	8.00	8.00
1974-75	8.50	8.50
1975-76	9.50	8.50
1976-77	9.50	8.50
1977-78	9.50	8.50
1978-79	10.00	10.00
1979-80	10.00	12.50
1980-81	13.00	13.00
1981-82	15.50	13.00
1982-83	15.50	13.00
1983-84	16.00	14.00
1984-85	16.00	14.00
1985-86	16.50	16.50
1986-87	--	17.00
1987-88	17.45	18.50
1988-89	19.00	19.50
1989-90	20.00	22.00
1990-91	23.00	23.00
1991-92	26.00	26.00
1992-93	29.00	31.00
1993-94	36.50	34.50
1994-95	39.50	39.10
1995-96	42.50	42.50
1996-97	45.90	45.90

Source: Sugar Year Book, 1983-84 and 1997-98.

PRICE OF SUGARCANE RECOMMENDED BY CACP AND SMP

From the table 5, it is clear that there are variations in the price recommended by Agricultural Price Commision (CACP) and SMP fixed by the Government of India. For example, in the initial years i.e.from 1967-68 to 1970-71, the prices fixed by the government are slightly higher than the price recommended by the CACP. During the years 1975-76, 1976-77 and 1977-78, the prices fixed by the government were lower than the CACP prices. This might be due to bumper sugarcane crop during these years. Similar situation prevailed even during the years 1981-82 to 1984-85. After 1985-86, the situation was reverse as the SMP fixed by the Government were higher than the CACP price with the exception of years 1990-91, 1991-92, 1993-94 and 1994-95. In these years CACP prices were higher than the SMP fixed by the government.

CONCLUSIONS

The foregoing discussion leads us to draw the following conclusions:

1. India is the largest producer of sugarcane in the world. There has been considerable importance to this commercial crop during the last two de cades. In the year 2003-04, at all India level, the area under sugarcae was 39.95 lakh hectares, sugar production 2362 lakh tonnes, yield 59.1 tonnes per hectare, sugar production 135.46 lakh tonnes, sugar cane crushed 1325.11 lakh tonnes and sugar recovery was 10.22 per cent. There were 435 sugar factories working in India during 2003-04.
2. Maharashtra is one of the leading sugarcane growing states in India. It is a perennial irrigated crop grown in all the zones of the state. Maharashtra state occupied a well recognized place of pride on the sugarcane map of the country. In the year 2003-04, at all Maharashtra level, the area under sugarcane was 5.26 lakh hectares, sugarcane production 270 lakh tonnes, yield 51.30 lakh tonnes, sugarcane crushed 290.39 lakh tonnes and sugar recovery was 10.94 per cent. There were 136 sugar factories working in Maharashtra during 2003-04.
3. Overall average utilization of plant capacity in the Maharashtra state was between 85.83 per cent and 99.36 per cent.
4. There are variations in the price recommended by Agricultural Price Commission (CACP)and SMP fixed by the Government of India.In the initial years i.e. from 1967-68 to 1970-71, the prices fixed by the government are slightly higher than the price recomended by CACP.Similar situation prevailed even during the years 1981-82 to 1984-85. After 1985-86, the situation was reverse as the SMP fixed by the government were higher than the CACP prices with the exception of years 1990-91, 1991-92, 1993-94, and 1994-95.
5. On an average, the sugarcane price fixed by State Government was Rs. 335.43 per metric tonne during the study period. As well as the price fixed by the Central Government shows continuously raising trend during the study period. During the study period the price of sugarcane fixed by the central Government was, on an average,

Rs.288.03per metric tonne. The price fixed by central and state Governments showed a positive trend during the period from 1981-82 to 1999-2000.

NOTES AND REFERENCES

1. Noel Deer, History of Sugar,1949, Vol. I, p. 15, Champman and Hall Ltd., London.
2. Kalyankar S.P., Kosle P.V. and Nagargoje S.R., Contribution of Yield and Area to Increase Sugarcane Production in Maharashtra, Journal of Agricultural Economics, Vol. 7, No. 1, March, 1995.
3. Korrane V.M., Nagapure S.C. and Ulemale D.H., Growth performance of Sugarcane in Maharashtra State, Maharashtra Journal of Agricultural Economics, Vol. 7, No. 1, March, 1995.
4. Agrawal R.N., Sugar Industry of India, My Recollection, Popular Prakashan Pvt. Ltd., Bombay. 1976.
5. Bapat N.G., Economic Development of Ahmednagar District 1960-1981, Bombay Progressive corporation Pvt. Ltd., 1973.
6. Chandra Satish, Agricultural Price Policy in India, Chug Publication, Allahabad, 1985.
7. Inamdar N.R., Government and Cooperative Sugar Factories, Popular Prakashan, Bombay, 1965.
8. Mohite V.J., Sugarcane Industry of Maharashtra. (A Blue Print for Progress), Government of Maharashtra, Bombay, 1974.

AMBHORE S.S.

DAIRY DEVELOPMENT IN MAHARASHTRA

10

INTRODUCTION

Agriculture is considered as the backbone of the Indian economy. With a 24.2 per cent contribution (triennium ending 2001-02) to the gross domestic product (GDP), agriculture still provides livelihood support to about two-thirds of country's population. The sector provides employment to 56.7 per cent of country's work force and is the single largest private sector occupation. Agriculture accounts for about 14.7 per cent of the total export earnings and provides raw material to a large number of industries (textiles, silk, sugar, rice, flour mills, milk products). Besides, the rural areas are the biggest markets for low-priced and middle-priced consumer goods, including consumer durables and rural domestic savings are an important source of resource mobilization. Any change in this sector, positive or negative, has a multiplier effect on the entire economy. A nation of more than a billion people cannot be dependent on imports for the basic item like food grains. The agriculture sector, therefore, acts as a bulwork in maintaining food security and, in the process, national security as well.

Dairy farming is also very important subsidiary occupation. Dairing is one of the most effective instruments for supplementing farmer's income and generating employment in the rural sector. It provides employment to millions of unemployed and under employed, and particularly to small farmers, marginal farmers and landless labourers in villages. Dairing is being envisaged as an important means of reinforcing the income for economically weaker section of the society constituted by landless and small farmers. Milk constitutes the most important source of nourishment for both vegetarian and non-vegetarian, for old and young alike. Its importance to the consumers is next only to that of their staple food.

The planners of the country consider cooperation as an instrument of economic development of the disadvantaged particularly in the rural areas. The features of cooperatives like non-exploitativeness, voluntary membership, principle of one person one vote, decentralised decision making and self-imposed curbs on profit have eminently qualified it an important instrument of development. The continuous strategic and planned efforts for the healthy growth of this movement in the country have resulted in spectacular progress of cooperative movement in the recent years.

The cooperative dairy movement in Maharashtra has played a vital and dominant role in the social and economic development of the state particularly in the rural areas. This has resulted in raising the standard of living of the rural people. This paper explains the main aspects or dairy development in Maharashtra.

Table 1
Population of Cattle, Buffaloes, Sheep and Goats in Maharashtra

(In thousands)

Year	*Cattle*	*Buffaloes*	*Sheep*	*Goats*	*Total*
1951	13.565 (61.68)	2.604 (11.84)	2.037 (9.26)	3.785 (17.21)	21.991 (100)
1956	14.109 (60.28)	2.721 (11.63)	1.970 (8.42)	4.606 (19.68)	23.406 (100)
1961	15.328 (59.67)	3.087 (12.02)	2.093 (8.15)	5.181 (20.17)	25.689 (100)
1966	14.729 (58.69)	3.042 (12.12)	2.205 (8.79)	5.121 (20.40)	25.097 (100)
1972	14.705 (56.46)	3.301 (12.67)	2.128 (8.17)	5.911 (22.70)	26.045 (100)
1978	15.218 (51.90)	3.899 (13.30)	2.636 (8.99)	7.563 (25.80)	29.316 (100)
1982	16.162 (52.97)	3.972 (13.03\2)	2.671 (8.75)	7.705 (25.25)	30.510 (100)
1987	16.983 (50.24)	4.755 (14.07)	2.873 (8.50)	9.195 (27.20)	35.806 (100)
1992)	17.441 (48.58)	5.447 (15.17)	3.074 (8.56)	9.941 (27.69)	35.903 (100)
1997	18.047 (50.27)	6.816 (18.98)	3.162 (8.81)	11.323 (31.54)	39.348 (100)
Quinquennial Growth rate (%)	4.17	16.13	6.87	16.64	8.80

Note: Bracketed figures indicate per centage to total.
Source: All India Livestock and Farm Equipment Census 1982 to 1997, Office of the Commissioner, Department of Animal Husbandry, Maharashtra State, Pune.

LIVESTOCK POPULATION

The data about livestock population in Maharashtra are presented in table 1. This table indicates that the livestock population in the state has grown at the quinquennial rate of 8.80 per cent. This growth rate was very high in case of goats (16.64 per cent) and buffaloes (16.13 per cent). The cattle population grew at the quinquennial rate of 4.17 per cent and sheep population at the rate of 6.97 per cent between 1951-97. At the advent of the first five year plan (1951), the livestock population of the state consisted of 61.68 per cent cattle, 17.21 per cent goats, 11.84 per cent buffaloes and 9.26 per cent sheep. This composition of the livestock population in the state has undergone a significant change over the period. According to 1997 livestock census, the total population of the state consisted of 50.27 per cent cattle, 31.54 per cent goats, 18.98 per cent buffaloes and 8.81 per cent sheep. It means that the proportion of cattle population has substantially declined whereas the proportion of goat population has increased very much in Maharashtra during the period under consideration. The proportion of buffalo population has shown a significant increase from 11.84 per cent in 1951 to 18.98 per cent in 1997. The proportion of sheep population has declined marginally over the period under consideration.

HEALTH FACILITIES FOR LIVESTOCK

The data about health facilities for animals in Maharashtra for the period 1970-71 to 1999-2000 are presented in table 2. It is evident from the table that the number of veterinary hospitals was 26 in 1970-71, which went up to 56 in 1980-81. But there were only 31 veterinary hospitals in the state since 1999-2000, which was due to the policy of the Government of Maharashtra. There has been substantial increase in the number of veterinary dispensaries from 263 hospitals in 1970-71 to 1209 (a little over three and half-fold increase) hospitals in 1999-2000.

Table 2
Health Facilities for Animals in Maharashtra (1970-71 to 1999-2000)

(Numbers)

Year	*Veterinary Hospitals*	*Veterinary Dispensaries*	*Artificial Insemination Centers*	*Cattle Breeding Farms*
1970-71	26	263	870	15
1974-75	55	241	1456	14
1980-81	56	507	2417	12
1985-86	56	764	2769	13
1990-91	31	815	3602	12
1991-92	31	920	3651	13
1992-93	31	924	3841	13
1995-96	31	1090	4067	12
1998-99	31	1156	4249	12
1999-00	31	1209	4453	12

Sources: 1. Office of the Commissioner of Animal Husbandry, Government of Maharashtra, Pune.
2. Statistical Abstract 1970-71 to 1999-2000, Directorate of Economics and Statistics, Planning Department, Government of Maharashtra, Mumbai.

The number of artificial insemination centers has gone up substantially (a four-fold increase) from 870 centers in 1970-71 to 4,453 centers in 1999-2000. The number of cattle breeding farms showed a marginal decline during the period under consideration. There were 15 farms in 1970-71 which declined to 12 farms in the state in 1999-2000.

GROWTH OF DAIRY CO-OPERATIVES IN MAHARASHTRA

The data about number and membership, capital and borrowings of cooperative dairy unions in Maharashtra during 1988 to 1997 are presented in table 3. It shows that the number of cooperative dairy unions in Maharashtra was 72 in 1988, which showed a general downward trend over the period of ten years and became 63 in 1997. Because of this the annual average growth rate of number of cooperative dairy unions is negative (-1.40 per cent).

However, the membership of the cooperative dairy unions in Maharashtra has shown an upward trend during the period under consideration. The membership of cooperative dairy unions in the state was 16,800 in 1988, which went up continuously, with the exception of 1995, and reached to 42,900 in 1997. The membership of cooperative dairy unions had grown at the annual average rate of 18.86 per cent during the period under consideration.

Table 3
Growth of Dairy Co-Operatives in Maharashtra (1988 to 1997)

(Rs. In Lakh)

Year	*No. of Unions*	*Member ship (00')*	*Share Capital*	*Of which Govt.*	*Owned funds*	*Outstanding Borrowings*	*Working capital*
1988	72	168	767	110	3,217	3,122	11,666
1989	68	142	965	145	3,731	3,506	13,087
1990	69	150	1,164	148	3,723	4,683	13,450
1991	71	243	1,166	159	4,272	3,868	22,663
1992	67	260	2,325	159	4,960	4,139	29,356
1993	67	269	5,816	159	9,003	4,562	32,313
1994	67	292	3,737	162	7,782	4,878	35,201
1995	62	164	3,458	238	8,782	4,826	23,909
1996	64	373	3,454	150	7,899	12,755	31,615
1997	63	429	5,226	326	13,723	20,593	26,555
Annual Growth Rate (%)	-1.40	18.86	33.79	18.94	21.20	30.81	12.90

Source: Cooperative Movement at a Glance in Maharashtra 1988 to 1999, Office of the Commissioner of Co-operation and Registrar Cooperative Societies, Maharashtra State, Pune.

It seems from the table that the share capital, owned funds, working capital and borrowings outstanding have shown a rising trend during the period under consideration. The total share capital of the cooperative dairy unions has grown at the annual average rate of 33.79 per cent, and at the same time government's share in total share capital of the cooperative dairy unions has grown at the annual average rate of 18.94 per cent. Owned funds of the cooperative dairy unions in the state have recorded a positive growth rate of 21.20 per cent per annum during the period under consideration. Borrowings outstanding of the cooperative dairy unions have also grown at a very high rate of 30.81 per cent per annum. Working capital of the unions has exhibited a positive growth rate of 12.90 per cent per annum during the period. From the above it can be said that the cooperative dairy unions in the state have done well on the front of membership, share capital, owned funds and working capital during the period under consideration.

MILK PRODUCTION

The table 4 shows that the milk production in Maharashtra State has grown at the annual average rate of 6.01 per cent during 1970-71 to 2000-01. The milk production in the state grew at the highest annual rate of 7.25 per cent during the eighties. It grew at the rate of 5.22 per cent per annum during the seventies and at the annual rate of 5.53 per cent during nineties.

Table 4
Growth Rate of Milk Production in Maharashtra

Period	*Growth rate (Per cent)*
1970-71 to 1979-80	5.22
1980-81 to 1989-90	7.25
1990-91 to 2000-01	5.53
1970-71 to 2000-01	6.01

Source: Based on data collected from office of the Dairy Development Department, Government of Maharashtra, Mumbai.

REGIONWISE TOTAL MILK PRODUCTION IN MAHARASHTRA (1992-93 TO 2001-01)

The data about regionwise total milk production in Maharashtra during 1992-93 to 2000-01 are presented in table 5. The table indicates that the total milk production in Maharashtra has shown a steady increase during the aforesaid period. The estimated milk production in the state has grown at the annual average rate of 4.61 per cent. Among the different regions of the state, the estimated milk production has recorded the highest rate of 6.36 per cent in case of Nasik division followed by Pune division (5.80 per cent). The growth in milk production in these two regions was comparatively higher than that at the state level. The estimated milk production in Mumbai (1.52 per cent), Amaravati (2.84

Table 5

Regionwise Total Milk Production in Maharashtra (1992-93 to 2001-01)

(Lakh Kg.)

Region / State	*1992-93*	*1993-94*	*1994-95*	*1995-96*	*1996-97*	*1997-98*	*1998-99*	*1999-00*	*2000-01*	*Annual Growth Rate (%)*
Mumbai	5,048.88 (12.31)	5,014.33 (11.80)	4,789.35 (9.95)	5,168.47 (10.15)	5,248.24 (10.24)	5,295.94 (10.20)	5,607.69 (10.00)	5,606.04 (9.82)	5,668.30 (9.82)	1.52
Nasik	8,830.94 (21.53)	9,219.37 (21.69)	11,475.87 (23.85)	12,171.69 (23.90)	12,135.45 (23.67)	12.254.16 (23.60)	13,348.73 (23.80)	13,748.66 (24.09)	14,206.95 (24.28)	6.36
Aurangabad	7,938.61 (19.36)	8,219.99 (19.34)	8,841.08 (18.38)	9,319.85 (18.30)	9,237.94 (18.02)	9,402.74 (6.72)	10,076.42 (17.97)	10,178.73 (17.84)	10,341.81 (17.68)	3.40
Amravati	3,189.35 (7.78)	3,281.62 (7.72)	3,166.16 (6.58)	3,399.71 (6.68)	3,453.94 (6.74)	3,490.70 (6.72)	3,810.62 (6.79)	3,885.18 (6.81)	3,969.34 (6.79)	2.84
Nagpur	2,527.03 (6.16)	2,621.68 (6.17)	2,841.39 (5.91)	2,898.36 (5.69)	2,953.33 (5.76)	2,985.99 (6.72)	3,243.82 (5.78)	3,307.58 (5.80)	3,398.97 (5.81)	3.81
Maharashtra	41,012.16 (100)	42,503.88 (100)	48,114.41 (100)	50,917.34 (100)	51.264.19 (100)	51,920.26 (100)	56,085.04 (100)	57,061.92 (100)	58,499.28 (100)	4.61

Note: Bracketed figures indicate per centage to total.

Source: Animal Husbandry Statistics Reports 1992-93 to 2000-01, Commissioner, Department of Animal Husbandry, Government of Maharashtra, Pune

per cent), Aurangabad (3.40 per cent) and Nagpur (3.81 per cent) division has grown at the annual rates which are lower than the growth rate of 4.61 per cent per annum for the Maharashtra state. From this it can be concluded that the estimated milk production in the state has increased substantially in all the regions during the period under consideration. Further, Nasik and Pune divisions have recorded relatively higher growth in milk production.

MILK DISTRIBUTION

The data about the per day distribution of milk in rural area, distribution through Mumbai Milk Scheme and Mahananda Dairy in Maharashtra during 1991-92 to 1998-99 are shown in table 6. It could be seen from the table that the sale of milk per day by Mumbai Milk Scheme has grown at a very low rate of 1.59 per cent per annum during the period under consideration. The per day milk sold by the Mahananda Dairy grew at the annual average rate of 9.29 per cent during the same period. But the important thing is that the per day rural sale of milk by the dairy cooperatives has grown at a very high rate of 36.92 per cent per annum during the period under consideration. It means that the cooperative dairies in Maharashtra have performed well with respect to the distribution of milk in rural area of Maharashtra.

Table 6

Distribution of Milk Per day in Maharashtra (1991-92 to 1998-99)

(Lakh Litres)

Year	*Mumbai Milk Scheme*	*Local Rural Sale of Milk*	*Milk Distribution by Mahananda Dairy*
1991-92	6.73	9.35	4.06
1992-93	6.31	2.51	4.04
1993-94	5.62	2.18	4.87
1994-95	5.40	2.00	5.50
1995-96	5.72	2.87	5.70
1996-97	6.82	4.20	6.78
1997-98	6.85	3.81	7.59
1998-99	7.30	14.19	7.40
Annual Growth Rate (%)	1.58	36.92	9.29

Source: Handbook of Basic Statistics 1991-92 to 1998-99, Office of the Commissioner for Cooperation and Registrar of cooperative Societies, Maharashtra State, Mumbai.

MILK PROCUREMENT

The regionwise data about milk procurement in Maharashtra during 1992-93 to 2000-2001 are presented in table 7. This table shows that the milk procurement in the state has grown at the annual average rate of 4.95 per cent during the period under consideration. The procurement of milk has not grown at similar rates in all the regions of the state. The

Table 7
Regionwise Milk Procurement in Maharashtra (1992-93 to 2000-01)

(Lakh kg)

Region / State	*1992-93*	*1993-94*	*1994-95*	*1995-96*	*1996-97*	*1997-98*	*1998-99*	*1999-00*	*2000-01*	*Annual Growth Rate (%)*
Mumbai	68.00 (0.70)	88.00 (0.77)	72.00 (0.61)	60.00 (0.51)	68.00 (0.53)	63.80 (0.47)	101.32 (0.75)	73.20 (0.49)	69.35 (0.49)	3.44
Nasik	2,810.00 (29.09)	3,831.00 (33.43)	3,214.00 (27.19)	3,070.00 (25.94)	3,381.00 (26.59)	3,643.61 (26.99)	3,647.37 (26.99)	3,773.46 (26.83)	3,679.20 (26.18)	4.34
Pune	5,250.00 (10.22)	5,676.00 (49.52)	6,511.00 (55.08)	6,844.00 (57.82)	7,242.00 (56.97)	7,568.68 (56.33)	7,726.28 (57.18)	7,989.78 (56.80)	8,048.25 (57.26)	5.56
Aurangabad	987.00 (10.22)	1,081.00 (9.43)	1,409.00 (11.92)	1,347.00 (11.38)	1,302.00 (10.24)	1,365.90 (10.16)	1,418.82 (10.50)	1,632.36 (11.61)	1,781.20 (12.67)	8.13
Amravati	158.00 (1.64)	278.00 (2.43)	199.0 (1.68)	113.00 (0.95)	120.00 (0.94)	135.09 (1.01)	63.79 (0.47)	98.82 (0.70)	62.05 (0.44)	-1.50
Nagpur	386.00 (4.00)	507.00 (4.42)	415.00 (3.51)	403.00 (3.40)	600.00 (4.72)	660.43 (4.91)	555.36 (4.11)	497.76 (3.54)	416.10 (2.96)	3.32
Maharashtra	9,659.00 (100)	11,461.00 (100)	11,820.00 (100)	11,837.00 (100)	12.713.00 (100)	13.437.51 (100)	13,512.54 (100)	14,065.38 (100)	14,056.15 (100)	4.95b

Note: Bracketed figures indicate per centage to total.
Source: Animal-Husbandry, Statistics Reports 1992-93 to 2000-01, Commissioner, Department of Animals Husbandry, Government of Maharashtra, Pune.

milk procurement activity has expanded at the highest annual rate of 8.13 per cent in Aurangabad division followed by Pune division (5.56 per cent) during the period under study. On the contrary, the milk procurement has recorded a negative growth rate of –1.50 per cent in case of Amaravati division during the same period. The milk procurement activity has also grown in Nasik (4.34 per cent), Mumbai (3.44 per cent) and Nagpur (3.32 per cent) divisions as well. However, the growth rates found for these regions were comparatively lower than the rate of growth for the state as whole. The above discussion allows us to say that the milk procurement activity has grown substantially in Aurangabad and Pune divisions, and it has eroded in Amravati division during the period under study.

The table 7 shows that the per centage share of milk procured was the highest in case of pune division (54.35 per cent) followed by Nasik (29.09 per cent) division during the year 1992-93. It was very low in case of Mumbai (0.70 per cent), Amaravati (1.64 per cent), Nagpur (4.00 per cent) and Aurangabad (10.22 per cent) division in 1992-93. In 2000-01, the per centage share of milk procured was the highest again in case of Pune division (57.26 per cent), and next to this for Nasik (26.18 per cent), division. It was very low for Amaravati (0.44 per cent), Mumabi (0.49 per cent), Nagpur (2.96 per cent) and Aurangabad (12.67 per cent) divisions. This indicates that the per centage share in total milk procurement has improved in case of only Pune and Aurangabad divisions in 2000-01 over 1992-93. From the above it is evident that the milk procurement activity is largely concentrated in Pune, Nasik and Aurangabad divisions and these divisions have improved their share in state's total milk procurement.

SEASONAL PATTERN OF MILK PROCUREMENT

The monthly data about total milk procurement by the cooperatives and government in Maharashtra for the period from 1992-93 to 2000-01 are presented in table 8. It could be seen from the table that the total milk procurement in Maharashtra was maximum in the month of February in five years out of nine years. During two years, it was maximum in the month of December and in. It was maximum in march considering the whole period together, one year it seems that the average per day milk procurement in Maharashtra is found to be maximum in the month of February. The minimum milk procurement was found in the month of July during the first four years and in the month of August during the three years. Considering the entire period of eight years together, it is found that the per day milk procurement was the minimum in the month of July. The monthly variation in total milk procurement in Maharashtra is found to be the highest during 1995-96 as both range between maximum and minimum milk procurement (8.77 lakh litres

Table 8

Seasonal Pattern of Milk Procurement by Co-operative and Government in Maharashtra. (1992-93 to 2000-01)

Lakh litre/Day)

Months	*1992-93*	*1993-94*	*1994-95*	*1995-96*	*1996-97*	*1997-98*	*1998-99*	*1999-00*	*2000-01*	*Average*
January	26.62	32.20	34.36	37.33	37.33	39.65b	37.41	38.30	41.79	36.11
February	26.95	33.15b	34.72b	37.39	37.39	39.46	37.61	38.95	41.97	36.40b
March	28.28	32.77	34.02	37.03	37.03	38.64	37.46	38.27	41.97b	36.17
April	23.65	28.68	31.03	32.97	32.97	35.90	36.45	37.39	36.68	32.86
May	23.76	28.13	30.15	30.78	30.78	34.45	34.80	36.28	35.88	31.67
June	23.09	27.57	28.31	29.47	29.47	32.76	34.00a	35.83	35.51	30.67
July	22.37a	27.29a	26.97a	28.99a	28.99a	31.47	37.50	35.83	35.23	30.18a
August	24.63	28.66	27.54	30.69	30.69	31.46a	34.23	33.31a	34.36a	30.62
September	27.67	30.39	29.52	32.35	32.35	33.98	34.54	36.07	36.84	32.63
October	28.49b	32.35	32.30	34.88	34.88	35.79	35.72	37.83	39.41	34.63
November	27.62	33.12	34.22	37.76b	37.76b	37.80	38.06b	38.93	41.81	36.34
December	27.03	32.11	33.59	37.01	37.01	38.40	37.88	39.45b	40.88	35.93
Average	25.85	30.54	31.39	33.89	33.89	35.81	36.06	37.20	38.53	33.68
Range Bet. Max & Min	6.12	5.86	7.75	8.77	8.77	8.19	4.06	6.14	7.61	6.22
S.D.	2.10	2.22	2.73	3.24	3.24	2.87	1.52	1.71	2.92	2.40
C.V.	8.13	7.27	8.70	9.57	9.57	8.02	4.21	4.58	7.57	7.12

a = Min. Level S.D. = Standard Deviation
b = Max. Level C.V. = Co-efficient of variation
Source: Office of the Commissioner, Department of Dairy Development, Government of Maharashtra, Mumbai.

per day) and coefficient of variation (9.57 per cent) were the highest for this year.

On the other hand, the monthly variation in total milk procurement is found to be the lowest during 1998-99 as indicated by the range between maximum and minimum milk procurement (4.06 lakh ltres per day) and coefficient of variation (4.21 per cent). From the above it can be said that total milk procurement in Maharashtra was the maximum in either February or December whereas it was minimum in either July or August during the period under consideration . Further, seasonal milk procurement in Maharashtra was most stable in 1998-99 and it was most unstable during 1995-96.

MILK PRICE

The data about milk purchase price by the cooperative dairy societies in Maharashtra are given in table 9. The cow milk price offered by the cooperative societies was Rs. 2.50 in flush season and Rs. 3.00 in lean season as on August 1983 in Maharashtra.

Since then the prices of cow milk offered by the societies have been upwardly revised and it were Rs. 7.50 in flush season and Rs. 8.50 in lean season since September 4, 2002, in Maharashtra state. It seems that there has been three-fold increase in flush season price of cow milk in Maharashtra. The lean season price of cow milk offered by the cooperative societies has also shown almost three-fold increase (283.33) during the same period.

Prices offered by the cooperative societies for buffalo milk during the flush and lean seasons have also shown a continuously rising trend between August 1, 1981 and September 4, 2000 in Maharashtra state. There has been a little over three-fold increase in the flush season and lean season prices of buffalo milk offered by the societies in the state. It is obvious that in case of both cow and buffalo milk, the lean season price has been higher than the flush season price throughout the period under consideration. It is to be noted that the per litre commission paid to the primary cooperative dairy societies has shown a little over four and half-fold increase over the period under consideration.

From the above it can be concluded that the cow and buffalo milk prices offered by the cooperative dairy societies of the Maharashtra state have indicated three-fold increase while the per litre commission of the cooperative dairy societies recorded a little over four and half-fold increase during the period under consideration.

SALE PRICE OF MILK

The data about sale prices of cow and buffalo milk charged by the cooperative dairy sector in Maharashtra are Shown in table 10. The prices of cow and buffalo milk charged by the cooperative dairies have

Table 9
Purchase Price of Milk in Maharashtra

Period	*Cow Milk Price*				*Buffalo Milk Price*				*Per Litre Commi-ssion of the Societies*	*Index of Commi-ssion*
	Flush Season (Rs.)	*Nominal Price Index*	*Lean Season (Rs.)*	*Nominal Price Index*	*Flush Season*	*Nominal Price Index*	*Lean Season (Rs.)*	*Nominal Price Index*		
August 1,81 to April 26,86	2.50	100.00	3.00	100.00	2.90	100.00	3.40	100.00	0.28	100.00
April 27,83 to February 28,88	3.00	120.00	3.40	113.33	3.70	127.59	4.18	122.94	0.38	135.71
March 1,88 to August 14,88	3.40	136.00	3.90	130.00	4.18	144.14	4.80	141.18	0.38	135.71
August 15,88 to March 22,91	3.60	144.00	4.50	150.00	4.43	152.76	5.33	156.76	0.38	135.71
March 23,91 to April 30,92	4.50	180.00	5.50	183.33	5.50	189.66	6.80	200.00	0.52	185.71
May 1,92 to October 31,95	5.50	220.00	6.70	223.33	6.80	234.48	8.00	235.29	0.75	267.86
November 1,95 to April 30,98	6.70	268.00	7.70	256.67	7.80	268.97	9.00	264.71	0.75	267.86
May 1,98 to December 31,2000	7.70	308.00	8.70	290.00	8.80	303.45	10.00	294.12	9.95	339.29
January 1,2001 to September 3,2002	8.50	340.00	9.50	316.67	9.60	331.03	10.80	317.65	1.30	464.29
September 4, 2002 to this date	7.50	300.00	8.50	283.33	9.60	331.03	10.80	317.65	1.30	464.29

Source: Office of Commissioner, Dairy Development Department of Maharashtra, Mumbai.

also increased over the period under consideration. The cow milk price charged by dairies was Rs. 3.90 per litre during April 27, 1983 to February, 1988 in Maharashtra. The price of buffalo milk charged by the dairies was Rs. 5.60 during the same period. The sale prices of cow and buffalo milk have been upwardly revised from time to time and they are Rs. 12.00 And Rs. 16.00 Per litre from January 1, 2001. There has been 207.66 per cent and 185.71 per cent increase in the sale price of cow and buffalo milk respectively in Maharashtra during the period under consideration.

Table 10
Sale Price of Milk in Rural Maharashtra

(Rs./Litre)

Period	*Buffalo Milk Price*	*Cow Milk Price*
April 27, 83 to February 28, 88	5.60	3.90
March 1,88 to August 14, 88	7.00	5.00
August 15,88 to March 22,99	7.00	5.00
March 23, 91 to March 30, 92	9.00	6.50
May 1,92 to October 31,95	11.00	8.00
November 1,95 to May 19,97	11.00	8.00
May 20,97 to January 14,2000	15.00	11.00
January 15, 2000 to December 31, 2000	16.50	12.50
January 1, 2001 to this date	16.00	12.00

Source: Office of the Commissioner of Dairy Development, Government of Maharashtra, Mumbai.

MILK PROCESSING

The table 11 shows the average collection of milk, number of milk plants and their capacity and number of milk chilling centers and their capacity in Maharashtra. It is evident from the table 11 that the milk collection in the state has steadily increased from 17.42 lakh litre in 1984-85 to 37.37 lakh litre in 1999-2000. It means there has been 114.52 per cent increase in milk collection in the state over the period under consideration. The number of milk plants was 25 in 1984-85 which increased to 33 in 1990-91, declined to31 in 1995-96 and again increased to 32 in 1997-98. Their milk processing capacity was 28.25 lakh litres in 1984-85 which increased to 42.40 lakh litres in 1990-91, declined to 32.10 lakh litres in 1995-96 and then slightly increased to 32.30 lakh litres in 1997-98. The number of milk chilling plants has shown a slow declining trend during the period under consideration. It was 83 in 1984-85, which declined steadily to 69 in 1998-99. Similarly, their milk chilling capacity has also steadily declined from 8.58 lakh litre in 19984-85 to 6.57 lakh litre in 1998-99.

The table 12 shows the number of milk chilling plants and their capacity with cooperative and government dairies in Maharashtra. This table indicates that both the number and capacity of chilling plants with the cooperative dairies has grown duirng 1989-90 to 1998-99. The number of plants increased at the rate of 4.23 per cent per annum while their processing capacity has grown at the annual average rate of 11.80 per cent during the period under consideration. The number of milk chilling plants with the government dairies has remained almost stagnant but their chilling capacity increased at a very high annual rate of 23.71 per cent during 1991—92 to 1998-99.

Table 11
Number of Milk Plants and Milk Chilling Centers and their Capacity in Maharashtra (1984–2000)

Year	*Average milk collection (in Lakh litres)*	*Milk Plant*		*Milk Chilling Centers*	
		Number	*Capacity (in lakh litres)*	*Number*	*Capacity (in lakh liters)*
1984-85	17.42	25	28.25	83	8.58
1990-91	23.78	33	42.40	82	8.16
1995-96	31.49	31	32.10	77	7.19
1996-97	33.89	31	32.10	77	7.19
1997-98	35.56	32	32.30	80	7.64
1998-99	36.02	32	32.30	69	6.57
1999-00	37.37	32	32.30	69	6.57

Source: Office of the Commissioner of Dairy Development, Government of Maharashtra, Mumbai.

Table 12
Government and Cooperative Dairies and their Processing Capacity in Maharashtra (1989-90 to 1998-99)

Year	*Cooperative Dairies*		*Government Dairies*	
	Number of Milk Processing Plants	*Capacity (in lakh litres)*	*Number of Milk Processing Plants*	*Capacity (in lakh litres)*
1989-90	18	12.15	N.A	N.A
1990-91	18	12.15	N.A	N.A
1991-92	24	12.30	32	12.15
1992-93	24	13.40	32	32.65
1993-94	25	14.90	31	33.15
1994-95	25	19.70	30	31.40
1995-96	25	23.25	31	31.80
1996-97	25	23.25	31	32.10
1997-98	27	25.75	32	31.30
1998-99	25	31.90	32	31.70

Source: Office of the Commissioner of Dairy Development, Government of Maharashtra, Mumbai.

PRODUCTION OF MILK PRODUCTS

The data about production of different milk products in Maharashtra during 1986-87 to 2000-01 are given in table 13. It is seen from the table that the production of ghee has fluctuated very much during the period under study. However, it indicated the growth rate of 3.38 per cent per annum in Maharashtra during the period under consideration. Butter production in Maharashtra has recorded a positive growth rate of 9.40 per cent per annum during the same period. The paneer production has also grown at the rate of 3.72 per cent per annum. The cheese production grew at the rate of 3.82 per cent per annum while the production of shrikhand recorded 7.55 per cent growth rate in Maharashtra during the same period.

Table 13
Production of Milk Products in Maharashtra (1986-87 to 2000-01)

Metric tonnes)

Year	*Ghee*	*Table Butter*	*Paneer*	*Cheese*	*Shrikhand*	*Skimmed Milk Powder*	*White Butter*
1986-87	1052.75	527.00	25.67	28.91	36.35	6931.00	4457.00
1987-88	1219.00	741.45	28.90	35.06	43.91	3376.28	1721.64
1988-89	1272.00	473.69	23.12	58.02	34.06	6924.87	3343.36
1989-90	1763.42	491.70	32.50	38.96	35.63	11567.69	5859.09
1990-91	1943.23	496.97	28.84	59.74	34.03	9904.61	4449.16
1991-92	1748.34	692.48	21.83	60.54	41.95	5892.63	2727.97
1992-93	1305.76	300.24	20.81	54.99	48.16	6581.42	3084.57
1993-94	2142.23	263.07	22.54	54.07	45.19	8493.52	4493.52
1994-95	2033.49	796.75	24.38	47.97	55.72	4877.20	7996.87
1995-96	1799.96	380.03	30.17	52.82	40.19	3719.08	1952.73
1996-97	1648.17	139.65	30.60	347.64	51.29	4208.40	2458.52
1997-98	1283.56	236.45	33.03	41.73	62.23	5638.28	3138.63
1998-99	12360.00	310.40	33.80	34.00	78.30	4688.80	2512.13
1999-00	1298.34	203.33	34.40	35.10	79.20	10651.15	6091.64
2000-01	1241.00	200.00	36.30	38.40	82.34	9934.00	5530.00
Annual Growth Rate (%)	3.38	9.40	3.72	3.82	7.55	13.65	19.52

Sources: 1. Commissioner of Dairy Development. Government of Maharashtra, Mumbai.
2. Handbook of Basic Statistics 1985 to 1999, Office of the Commissioner for Cooperation and Registerar of Cooperative Societies, Maharashtra State, Mumbai.

The production of skimmed milk powder and white butter have grown at the annual average rate of 18.65 per cent and 19.52 per cent respectively during the period under consideration. The above discussion reveals that the production of ghee, butter, paneer, cheese and shrikhand have recorded positive but low growth whereas skimmed milk powder and white butter production exhibited substantial growth during the period under consideration.

CONCLUSION

The main conclusions of the present study presented in the earlier section have indicated some strengths and weaknesses of the cooperative dairy sector in Maharashtra. The conclusions of the present study have the following policy implications:

1. The study indicated three important aspects of livestock population in the Maharashtra State First it is growing. Second, the composition of livestock population has undergone a significant change i.e. proportion of cattle population is declining whereas proportion of buffalo population is increasing. Third, the livestock population is largely concentrated in Ahemdnagar, Nasik, Pune, Parbhani, Yavatmal and Buldhana districts of the state. From this it can be suggested that attention needs to be given towards stopping the declining proportion of cattle population in the State. Because cattle power is still largely used as an input in agricultural production. Special efforts are required for promoting the sheep and goat rearing in the most backward districts of the State. Because this would help in raising the income level of the people in the backward districts of the Maharashtra state.
2. The study indicated that the facilities for health and improving the quality of livestock have increased in the State. Veterinary dispensaries and artificial insemination centers have grown substantially. However, more efforts are needed to improve the quality of cattle, buffaloes, sheep and goats in the State. Few more cattle breeding farms should be started in the districts where they are not available. The dairy farmers in Maharashtra do not have the facilities for improving the quality of indigenous buffaloes. They have to purchase the new breed buffaloes at high prices from the adjoining State of Gujrat. Hence it is necessary to start a few buffalo breeding farms in the State itself. There is also need of opening some cattle, sheep and goat breeding farms in the backward districts of the State.
3. The study of the growth of cooperative dairy institutions in Maharashtra indicated that the dairy societies/unions have performed well with respect to several parameters such as their number, membership, share capital, working capital, owned funds, etc. But government's share capital has declined and outstanding borrowings of the societies and unions have rapidly increased. It is, therefore, essential to pay sufficient attention towards reducing the outstanding borrowings of the dairy cooperatives in Maharashtra. Hereafter government will be withdrawing from the dairy sector and hence the societies will have to build up their own share capital without relying on the government.
4. The study indicated that the milk production in the State has grown at a substantially higher rate of 6.01 per cent during 1970-71 to 2000-01. The milk production from crossbred cows has grown at a

faster rate. Efforts are still needed to enhance cross bred cow and buffalo milk production in the State. Milk production is largely concentrated in Pune and Nasik divisions of the Maharashtra state, which are the most developed parts of the State. It means that the dairy business is at present more grown in these developed regions of the state. So the Government should promote dairy activity in other backward regions of the State as it creates employment and generates income for the rural families. The study revealed that Ahmednagar, Pune, Kolhapur, Satara and Jalgoan were the top milk producing districts whereas Gadchiroli, Chandrapur, Sindhurg, Ratnagiri and Wardha were the bottom milk producing districts of the Maharashtra State in 2000-01. This further reveals that dairy farming is mostly developed in the enconomically forward districts of the State. And the districts from Konkan and Vidarbha regions are lagging behind in dairy farming. It is, therefore, essential to promote the dairy farming in the backward districts of the State.

5. The study indicated that the number of cooperative dairy societies ate largely concentrated in Kolhapur, Ahmednagar, Pune, Solapur and Satara districts where dairy activity is developed. The cooperative dairy societies were lowest in number in Mumbai, Raigad, Gadchiroli, Ratnagiri and Thane districts of the State. It is, therefore, essential to open cooperative dairy societies in these districts.

6. The study indicated that the milk procurement in Maharashtra has increased at the highest annual rate in Aurangabad and Pune divisions but in Amravati division it showed a negative growth. The per centage share in milk procurement was the highest in case of Pune division followed by Nashik divisions during the year 2000-01. The per centage share in milk procurement was the lowest in case of Amaravati, Mumbai and Nagpur division during the year 2000-01. These results indicate that the milk procurement activity has expanded in those divisions of the State where milk production is high and network of cooperatives for milk procurement are well developed. In the backward divisions such as Amravati and Nagpur the procurement either declined or grown slowly due to lack of sufficient procurement network. It is therefore suggested that the sufficient network for milk procurement should be created on cooperative lines in the backward regions of the State.

7. The study also indicated that the milk collection by the government sector has grown at the annual average rate of 1.22 per cent whereas milk collection by cooperative sector has grown at the annual average rate of 8.96 per cent during the period 1992-93 to 2000-01. Though the procurement of milk through cooperatives has grown

substantially the existing network does not cover all the villages particularly remote villages in the State. Hence, there is need of extending the cooperative network for milk collection to all the villages in the State.

8. The study indicated that the number of milk processing plants was 32 and their capacity was 32.30 lakh litres per day in 1999-2000. The number of milk chilling centres was 83 in 1984-85 and 69 in 1999-2000. The capacity of the milk chilling centres was 8.58 lakh litres in 1984-85 and 6.57 lakh litres in 1999-2000. It means that the milk plants and their capacity have decreased during the study period. Hence there is need of increasing the milk processing and chilling facilities in the state.
9. The study indicated that the production of ghee, butter, paneer, cheese has grown in Maharashtra during the period from 1986-89 to 2000-01. Also the production of Shrikhand, Skimmed milk powder and white butter exhibited. positive growth during the same period. In this connection, it is suggested that the cooperative diary sector in Maharashtra should concentrate on the production of such milk products which are more price competitive in international market.
10. The study indicated that both the purchase and sale price of cow and buffalo milk have recorded three-fold increase but the commission of the agents recorded four and half-fold increase during the study period. The purchase of price of milk is fixed by the State government. The price should be fixed in such a manner that it would cover the total cost of milk production. The government should ensure remunerative price for the milk producers, which, in turn, would promote the diary business in the state.

REFERENCES

Economic Survey 2001-2002, Planning Department, Government of Maharashtra, Mumbai.

Maharashtra Co-operative, Aug. 2002. The Maharashtra State Cooperative Federation, Pune.

Indian Dairyman-Various Issues.

Daily News Paper Dainik Lokmat Dt. 26th Sept. 2002.

Daily News Paper Dainik Lokmat Dt. 5th Oct. 2002.

Economic Survey 2001-2002, Govt. of India, New Delhi.

Dairy Year Book 2002-2003.

Gupta P.R. Cindrella of Indian Agricultural, Manorama Year Book, 1998.

National Dairy Development Board, Annual Report 2001-2002.

Singh B. and Gill S.S. Milk production trends in India, Dairy Guide December 1986.

R.M. CHOUDHARI

PERFORMANCE OF PUBLIC DISTRIBUTION SYSTEM IN MAHARASHTRA

11

Banishing hunger and ensuring food security to all citizens has been accepted as the primary responsibility of the state towards its citizens, and is repeatedly endorsed at various national, regional and international flora. The concept of food security is interpreted in a variety of way. The concept of food security has been defined by the World Bank as ' Access by all people, at all times to enough food for an active, healthy life.' This is the widely used definition of food security. However, physical and economic access to food at the household level at all times to ensure healthy and active life is the crux of food security. Emphasis in this concept is on 'access' or 'entitlement' to individuals and not just food self-sufficiency at the national or household level. Individuals who do not get need-based food are defined as food insecure. In practice, food security is, generally, equated with the absence of hunger, or at best provision of pre-determined number of calories at the household level.

According to recent estimates, approximately 36 per cent of the Indian population lives below the poverty line. This means their income is not sufficient to buy enough food. About 80 per cent of these poor people live in rural areas. The Government of India as well as most State Governments have formulated and implemented various schemes to alleviate poverty or improve the social security of these most deprived people. The scheme's aim is to enhance self-employment, generate wage employment, or transfer income to the poor. The public Distribution System (PDS) is a large-scale food rationing programme meant to reduce food insecurity and improve welfare.

The origins of the Indian public distribution system can be traced back to the Second World War and the Bengal Famine in 1943. When the colonial government felt the need to develop a food policy for the country public distribution system was initiated in 1939 when perhaps for the first time Government intervened in the distribution of foodgrains in Bombay. The objective was mainly to check the speculative tendencies of grain merchants during the Second World War. Subsequently, the worst famine of the century (Bengal Famine of 1943), also underlined the urgency of PDS. Hence, the Government extended PDS to other states as well. By 1947, about 54 million people in urban areas were covered by statutory rationing and another 19 million by other forms of public distribution.

Public distribution comprised of two types of rationing viz., statutory rationing and informal rationing. The difference between these two types of rationing is that in the statutory rationed areas, the open market is legally barred from functioning and government undertakes the responsibility of supplying specific rationed quantities to consumers. In the areas of informal rationing, on the other hand, the Government meets only a portion of the food requirements of the consumers through fair price shops, and that too, only if the fair price shops have adequate quantities to offer. The open market is allowed to function so that the consumers can supplement the rations obtained from the fair price shops by purchases in the open market or, if they so desire, purchase their entire foodgrains requirement from the open market. Statutory rationing resorted to only in major cities, all other areas in the states being covered by informal rationing. The reason why statutory rationing has been instituted in major cities is that they are pockets of high purchasing power and if they are not cordoned off, prices will be pushed up all over the country.

The responsibility for operating the PDS is shared between the central and state Governments. The central Government procures, stocks and supplies grain and absorbs the costs of these operations. Once the grain is allocated to the states it is the job of the state Government to 'lift' the grain and distribute it to the ratail PDS outlets across the state. Hence, the PDS performance depends on foodgrains operations of the central government as well as the distribution of subsidized grain by state Governments. As a result, regional diversity in PDS performance can be expected and is indeed a finding of earlier studies.

During past decades, the Indian Government has subsidized the consumption of rice and wheat in most states through the public distribution system. The PDS is programme that involves arrangement between the central Government and state governments to distributive government procured grains through fair price shops. These shops are located primarily in urban areas. Eligible households can obtain ration cards for specified amounts of grain, and these are registered at the fair price shops. The Central Government determines procurement prices and quantities and the State Governments determine transportation charges and margins in arriving at issue prices, which are generally lower than open market retail prices. Evidence from several Indian studies indicate that the Indian government food subsidy policy helped poorer consumers through income redistribution.

The government operations till 1997 can be described as: higher procurement price, higher procurement, higher issue price, lower offtake and larger stocks. The introduction of targeting which created dual prices for above poverty line households and below poverty line households led to two opposing effects. The APL households, which had to pay an issue price equalto the economic cost of the FCI, moved out the the PDS reducing offtake. The BPL households received a substantial discount on the economic cost and had incentives to utilize the PDS. The states with a well - functioning adapted quickly to the targeted scheme and made full use of the BPL allocations. These are predominantly the rice consuming southern states of Andhra Pradesh, Tamilnadu and Kerala and they account for the bulk of the offtake. The principally wheat-consuming states of Rajasthan, Maharashtra and Uttar Pradesh which have weak public distribution systems were slow to identify BPL beneficiaries. As a result, wheat offtake fell drastically under the targeted PDS which rice offtake has remained at its earlier levels.

A serious objection to PDS has been voiced that due to PDS, prices of commodities in the open market would be higher than in situation where PDS is not operating. As a consequence, the poor are the net losers if they have to take recourse to the market for the bulk of their purchases. The weighted average price for purchase from PDS and from the open market would be unfavourable to them. The implication of the finding is not to dispense with PDS, but to enable the poor to obtain all, or nearly all, of the requirments of essential commodities from PDS and to rely on the open market to the least extent for such commodities. A corollary to this strategy is to discourse universal coverage by the PDS.

The new PDS system is however faced with several problems and doses not appear to address the real issues. It is argued that the cost effectiveness of the system appears to be low partly due to the fact that several indirect benefits of the system go unaccounted (price support to farmers, stability in prices of foodgrains, overall growth of the economy). Each of the economic costs of FCI, consisting of the purchase or procurement price, procurement incidentals, storage and distribution costs, has been growing at a faster rate and leading to rising food subsidy bill. The introduction of TPDS has only worsened the problem. TPDS appears to have failed in both of its objectives. It has succeeded neither in reducing the food subsidy bill nor in reaching food subsidies to a grater proportion of the poor.

The Government should now concentrate on tackling the real issues namely the cost inefficiencies in the procurement and distribution system and control the diversion of grain from PDS to the open Market. While some functions of the Food Corporation of India can be left to the market, there is a large scope for improving efficiency in the remaining operations of the Food Corporation of India. Geographical targeting holds substantial promise to bring the PDS effectively within the reach of a large number of the poor.

The Premier objective of this study is to evaluate the functioning of PDS in Maharashtra in general and Latur District of Maharashtra state in paricular, and assess the gains accrued to the beneficiaries under the system in relation to the costs.

This study used both secondary and primary data. Primary data were collected form 200 PDS bebeficiaries in Latur district of Maharashtra State in 1999-2000. There were ten talukas in the district during the survey year and 20 Card holders were selected at random form each taluka. form each taluka. Care was taken to wichide all types of card holders belonging to different social groups in the sample.

Latur district was created in 1982 through bifurcation of Osmanabad district. The five talukas out of eleven talukas belonging to old Osmanabad district were separated from it to form Latur district. These talukas were Latur, Ahmedpur, Udgir, Nilanga and Ausa. Latur on in 1992, a new taluka, namely Renapur, was created to include it in this new district. Previously, Renapur was a circle belonging to Ambajogai taluka of Beed district. At the same time a new taluka, namely Chakur, was created in this district. In June 1999, another three talukas viz., Devani, Jalkot and Shirur Anantpal were created. Hence since June 1999, there are ten talukas in this district.

The data about geographical area and administrative structure in Latur district in relation to Marathwada region and Maharashtra state are presented in table 1 below. It could be seen from this table that the total geographical area of the state is 308000 square kms. Out of this 2.33 precent geographical area is occupied by the Latur district. At present there are 373 talukas in Maharashtra state, of which 76 talukas (21.53 per cent) belong to Marathwada region and there are 10 talukas (2.83 per cent of the state) in Latur district. Out of 378 cities /towns in the state, 56 are in Marathwada region, of which only 7 are in Latur district. The number of inhabited villages in the state was 43722 in 2001, of which 15.52 per cent (or 6786 villages) were in Marathwada region and 2.11 per cent (or 923 villages) were in Latur district alone. Out of 224 municipal councils in the state, 51 were in Marathwada and only 5 in Latur district. The number of Panchayat Samities in the district was 7 as against 62 in Marathwada and 349 in the Maharashtra state. Out of 27832 gram panchyats in the state, 2.81 per cent (or 782) were in Latur district during 2001.

The following table shows the population and sex ratio of Latur district in comparison with Marathwada region and Maharashtra State.

This table shows that the population of the Maharashtra state was

Table 1
Area and Administrative Structure

		Latur District	*Marathwada region*	*Maharashtra State*
1.	Geographical area (sq. Km.)	7166	61811	308000
		(2.33)	(21.04)	(100)
2.	Talukas (No.)	10	76	353
		(2.83)	(21.53)	(100)
3.	Town/cities (No.)	7	56	378
		(1.85)	(14.82)	(100)
4.	Villages (1991) (No.)	945	85.23	43722
		(2.16)	(19.49)	(100)
5.	Inhabited villages (No.)	923	6786	43722
		(2.11)	(15.52)	
6.	Municipal Councils (No.)	5	51	224
		(2.23)	(22.77)	(100)
7.	Panchayat Samities (No.)	7	62	349
		(2.01)	(17.77)	(100)
8.	Gram Pranchayats (No.)	782	6631	27832
		(2.81)	(23.83)	(100)

Note: Figures in brackets indicate percentage to Maharashtra state.
Sources: 1. Aurangabad Division at a Glance 2000-01, Directorate of Economics and statistics, Goverment of Maharastra , Regional office, Auragbad, 2002-03.
2. Economic Survey of Maharashtra ,2002-03.

Table 2
Population, Urban-Rural Population and sex Ratio (2001)

		Latur District	*Marathwada region*	*Maharashtra State*
1.	Population (Lakhs)	20.78	155.89	967.562
	% to state	2.15	16.11	100
2.	Male population (Lakhs)	10.74	80.49	503.34
	% to State	(2.13)	(15.99)	(100)
3.	Female Population (Lakhs)	10.04	75.4	446.18
	% to State	(2.16)	(16.24)	(100)
4.	Decadal growth rate (1991-2001(%)	23.95	21.78	22.57
5.	Rural population (Lakhs)	15.88	117.57	557
	Percentage of rural population	76.42	75.42	57.6
6.	Urban population (Lakhs)	4.9	38.32	410
	Percentage of Urban pupulation	23.58	24.58	42.4
7.	Sex ratio (Females per thousand males)	934	936	922
8.	Density of Population (Per sq. km.)	290	240	314

Source: As indicated in table1.

9,67,52,000 in 2001. Out of which 155.89 lakhs (or 16.11 per cent) was in Maharathwada region and 20.78 lakhs or 2.15 per cent was in Latur district. Out of the total male population of the state 15.99 per cent was in Marathwada region and 2.13 per cent male population was in Latur district. Regarding female population, it was 2.16 per cent in Latur district and 16.24 per cent in the Marathwada region. From this it seems that the district's and region's share in state's female population was comparatively higher than their respective shares in state's male population in 2001. The current population census (of 2001) has indicated that decadal growth rate of population was relatively higher (23.95 per cent) in Latur district compared to Marathwada region level (21.78) per cent and state level (22.57 per cent) decadal population growth rates during 1991-2001. It means that population of Latur district has grown at a higher rate, when compared to region and state during the decade 1991-2001.

The food economy and public distribution system in India have shown a number of queer developments in recent years. It can be seen from table 3 that there has been year to year fluctuations in foodgrains output during the period 1990-2002. The foodgrains output had declined in 1991-92, 1995-96 and 2000-01 compared to their respective earlier years. The output of pulses hovered between 12-14 million tonnes. The per capita net availability of foodgrains (cereals and pulses) has declined considerably since 1998. But contrary to this, there has been a substantial increase in procurement of foodgrains. It was 22.81 million tonnes in 1990-91, which went up to 37.41 million tonnes in 2001-02. Similar trend is also observed in the procurement of rice and wheat individually.

Table 3
Production and Procurement of Foodgrains in India 1990-2002

Year	*Production (Million tonnes)*			*Procurement (Million tones)*		
	Cereals	*Pulses*	*Total foodgrains*	*Rice*	*Wheat*	*Total foodgrains*
1990-91	162.13	14.26	176.39	11.74	11.07	22.81
1991-92	156.36	12.02	168.38	9.24	7.75	16.99
1992-93	166.66	12.82	179.48	11.79	6.38	18.17
1993-94	170.95	13.31	148.26	13.65	12.83	26.48
1994-95	177.46	14.04	191.50	13.40	11.87	25.27
1995-96	168.11	12.31	180.42	9.95	12.33	22.28
1996-97	185.19	14.25	199.44	12.22	8.16	30.38
1997-98	179.28	12.98	192.26	14.33	9.30	23.63
1998-99	188.70	14.90	203.60	11.84	12.65	24.49
1999-00	196.40	13.40	209.80	17.27	14.14	31.40
2000-01	185.20	10.70	195.90	19.10	16.35	35.45
2001-02	197.80	13.50	211.30	16.78*	20.63	37.41

As on April 12, 2002.
Source: Economic Survey, Government of India,2002-03.

As far as the economic cost of food Corporation of India is concerned, the food e.g. wheat and rice are mentioned is the following table to show the cost.

The responsibility for operating the PDS is shared between the central and the state governments. The central government procures, stocks and supplies grain and absorbs the costs of these operations. Once

Table 4
Economioc Cost of Food Corporation of India

Year	*FCI's economic cost (Rs. / quintal)*		*FCI's economic cost as per cent of MSP*	
	Rice	*Wheat*	*Rice*	*Wheat*
1991-92	497	391	216.1	142.2
1992-93	585	504	216.7	152.7
1993-94	665	532	214.5	152.0
1994-95	695	551	204.4	153.1
1995-96	762	548	211.9	153.7
1996-97	858	663	225.8	139.6
1997-98	937	798	225.8	156.5
1998-99	995	800	226.1	145.5
1999-00	1111	872	226.7	150.3
2000-01	1148	830	225.1	136.1
2001-02	1174	839	225.5	135.3

Source: Economic Survey, Government of India, 2001-02 and 2002-03.

Table 5
Food Subsidy of the Central Government

(Rs. Crore)

Year	*food subsidy*	*Buffer component*	*Buffer component as per cent of food subsidy*
1996-97	6066	-	-
1997-98	7900	937	12.5
1998-99	9100	1552	17.8
1999-00	9434	1754	19.1
2000-01	12060	4233	35.2
2001-02	17612	5680	41.6
2002-03 (BE)	12200	-	-

Source: Economic Survey, Government of India, 2001-02 and 2002-03.

Table 6
Foodgrains Allocation and Offtake under Public Distribution System in India

(Million Tonnes)

Year	Wheat		Rice	
	Allocation	Offtake	Allocation	Offtake
1991-92	10.56	8.83	11.36	10.17
1992-93	9.25	7.85	11.48	9.55
1993-94	9.56	5.91	12.41	8.87
1994-95	10.80	4.83	13.32	8.03
1995-96	11.31	5.29	14.62	9.46
1996-97	10.72	8.52	15.10	11.14
1997-98	10.11	7.08	12.83	9.90
1998-99	10.11	7.95	12.93	10.74
1999-00	10.37	5.76	13.84	11.31
2000-01*	11.57	4.07	16.26	7.97
2001-02*	13.14	5.68	17.23	8.16
2002-03*	29.45	6.12**	27.35	7.39**

* Including Antyodaya ** Upto December
Source: Economic Survey, Government of India, 2001-02 and 2002-03.

the grain is allocated to the states, it is the job of the state government to lift the grain and distribute it to the retail PDS outlets across the state. Hence the PDS performance depends on foodgrains operations of the central government as well as the distribution of subsidized grain by state governments.

The data about quantity of rice and wheat allotted by the Government of India to the Maharashtra state are presented in table 4.14. This table indicates that the allocation of foodgrains by the central government to Maharashtra state has shown some decline during 1993 to 1997-98. During the remaining years of the period under consideration, allocation of foodgrains to Maharashtra state was higher compared to preceding year.

Table 7
Food Subsidy Under PDS in Latur District

(Rs. in lakhs)

Year	*Food subsidy*	*Percent change*
1997-98	97.22	-
1998-99	364.87	275.31
1999-00	661.05	81.18
2000-01	223.28	-66.22
2001-02	93.51	-58.12
2002-03	99.16	6.04

Source: District Civil Supplies Office, Latur.

Out of total fair price shopsin the state, 2.79 per cent are found in Latur district, in 2001. The number of FPSs inthe district has grown at the annual rate of 1.67 per cent as against the rate of 3.03 per cent at the state level, during 1990 to 2001. The supplies of rice and wheat to FPSs in the district have recorded a positive growth rate of 12.20 per cent and 18.88 per cent respectively in Latur district during the period 1990-91 to 2001-02. But the supply of sugar to FPSs in the district recorded a negative growth rate of -9.91 per cent per annum during the same period. The food subsidy was found to be very high during 1998-99 to 2000-01 in Latur district. Wheat first and then rice are the most important foodgrains distributed through PDS in the district under study. Allotted quota and offtake of both the grains have recorded substantial positive growth during the recent years.

Regarding the social status of the sample families, 24.80 per cent (or 62 families) were scheduled castes, 6.40 per cent (or 16 families) were scheduled tribes and 28.40 precent (or 71 families) were other backward castes. It means that 59.60 per cent of the sample families were from different backward classes. The remaining 40.40 per cent (or 101 families) were from different unreserved catagories.

Regarding the operation of fair price shops, several concerns are expressed in the literature and found true in empirical works. One of these concerns is the use of faulty weights and measures and underweighing of commodities at the time of sale by the FPS dealers. These practices might be followed by the FPS dealers just to earn more income from this business. In this connection, it is found that 87.20 per cent (or 218 out of 250) of the total sample PDS beneficiaries in the district under study, have said that the standard weights and measures are used for weighing commodities at the time of delivery. Only 32 sample families (or 12.80 per cent) have opined that the FPS dealers use faulty weights and measures and commodities was the maximum of 10.57 kg in Nilanga taluka and the minimum of 8.90 kg in Udgir taluka of the district under study. In case of sugar, the per family per month consumption requirement is found to be 6 kg. It varied in a narrow range of 6.45 kg in Devani taluka and 5.80 kg in Renapur taluka of the district. The per family per month requirement of kerosene is estimated at 3.86 litres in the district under study. It also varied in a narrow range across the talukas of the district. It was the maximum of 4.10 litres in Latur taluka and the minimum of 3.23 litres in Jalkot taluka of the district under study.

A majority of the beneficiaries draw goods several times in a month. They have to pay several visits to the FPSs for drawing the quota of goods. 81.20 per cent families were holding white card (BPL families), 93.60 per cent families were having ration shop in their own village, 84 per cent families have to walk a little distance to draw ration, 78 per cent families get ration reqularly from the FPS. These indicated the high access of beneficiaries to FPS in the study area. Delay in issue and renewal of ration cards and no display of price and stock positions were some of the irregularities in the functioning PDS in the study area. One-third of beneficiaries were facing the problem of queuing at shops, 1/4 th of the beneficiaries do notfind.

Form theforejoiong analysis it can be concluded that the PDS in India faces several problems such as mountiong buffer stocks of foodgrains, rising burden of food subsidy, poor offtake of foodgrains from PDS, etc. But the sample survey study canducted etc. Latur district of the state indicated that substantial income can be transfermed to poor families through the PDS to ensure food security and hance PDS needs to be continned in future also.

Regarding the operation of fair price shops, several concerns are expressed in the literature and found true in empirical works. One of these concerns is the use of faulty weights and measures and underweighing of commodities at the time of sale by the FPS dealers. These practices might be followed by the FPS dealers just to earn more income from this business. In this connection, it is found that 87.20

percent (or 218 out of 250) of the total sample PDS beneficiaries in the district under study, have said that the standard weights and measures are used for weighing commodities at the time of delivery. Only 32 sample families (or 12.80 percent) have opined that the FPS dealers use faulty weights and measures and commodities are under-weighed by the dealers at the time of delivery.

A majority of the beneficiaries draw goods several times in a month. They have to pay several visits to the FPSs for drawing the quota of goods. 81.20 percent families were holding white card (BPL families), 93.60 percent families were having ration shop in their own village, 84 percent families have to walk a little distance to draw ration, 78 percent families get ration reqularly from the FPS. These indicated the high access of beneficiaries to FPS in the study area. Delay in issue and renewal of ration cards and no display of price and stock positions were some of the irregularities in the functioning PDS in the study area. One-third of beneficiaries were facing the problem of queuing at shops, 1/4 th of the beneficiaries do notfind beneficiaries were facing the problem of queuing at shops, 1/4 th of the beneficiaries do notfind ration shops open during non-working hours and 4/5 th of the beneficiaries were in favour of continuation of PDS for the benefit of the poor.

The table 8 shows the per family purchases of different commodities from FPS's value of FPS purchases at PDS prices and open Market prices & commodity-wise gains from the FPS purchases in Latur district. Let us frist take up the issue of FPS purchases of essential commodities by the sample beneficiaries during the survey period. It seems from the table that the per family per month quantity of wheat purchased is found to be 17.83 kg. which forms 65.17 percent of the per family per month purchase of rice from FPSs is found to be 7.16 kg which is 75.45 percent of the per family per month requirement of rice, as indicated above, in Latur district. In case of rice, the per family per month requirement was found to be 6 kg. But none of the sample beneficiaries are found to have purchased sugar from the FPSs during the survey period, The commodity sugar has disappeared from the list of rationed goods distributed to economically weaker sections of the soceity. This happened due to a very marginal difference between the PDS price & open market price of sugar in the study area. The open market price of sugar is hovering around Rs.14 a kilo where as PDS retail price of sugar was Rs. 13 per kg. during the survey period. So the FPSs in the study area could not get arole in fulfilling the sugar requirement of sample beneficiaries due to a megare difference between the PDS retail price and prevailing open market price of sugar. In case of Kerosene the per family per month purchases by Sample beneficiaries from FPS, is found to be 3.5 litres which form 91.71 percent of the per family per month requirement, as indicated above, of this commodity in the district under study. From the

Table 8
Gains to Beneficiaries from PDS.

Taluka	*Per family per month quantity Purchased from PDS*				*per family value of purchases from PDS (Rs.)*				*Value of Purchases from PDS at market Price (Rs.)*				*per family gain from PDS (Rs.)*			
	wheat (Kg.)	*Rice (Kg.)*	*Sugar (Kg.)*	*Kerosene (Litre)*	*wheat (Kg.)*	*Rice (Kg.)*	*Sugar (Kg.)*	*Kerosene (Litre)*	*wheat (Kg.)*	*Rice (Kg.)*	*Sugar (Kg.)*	*Kerosene (Litre)*	*wheat (Kg.)*	*Rice (Kg.)*	*Sugar (Kg.)*	*Kerosene (Litre)*
1. Latur	18.00	7.37	-	3.67	82.40	47.17	-	44.04	144.00	81.07	-	51.38	61.20	33.90	-	7.34
2. Chakur	17.80	7.40	-	3.60	81.88	47.36	-	43.20	142.40	81.40	-	50.40	60.52	34.04	-	7.20
3. Jalkot	17.73	7.37	-	3.63	81.56	47.17	-	43.56	141.83	81.07	-	50.82	60.28	33.90	-	7.26
4. Renapur	17.95	7.50	-	3.45	82.57	48.00	-	41.40	143.60	82.50	-	48.30	61.03	34.50	-	6.90
5. Ahmedpur	16.23	6.97	-	3.50	74.66	44.61	-	42.00	129.84	76.67	-	49.00	55.18	32.06	-	7.26
6. Nilanga	16.03	7.47	-	3.63	73.74	47.81	-	43.56	128.24	82.17	-	50.82	54.50	34.36	-	7.26
7. Ausa	19.45	6.80	-	3.35	89.47	43.52	-	40.20	155.60	74.80	-	46.90	66.13	31.28	-	6.70
8. Shirur (A)	21.00	6.95	-	3.50	96.60	44.48	-	42.00	168.00	76.45	-	49.00	71.40	31.97	-	7.00
9. Devani	19.55	7.45	-	4.15	89.93	47.68	-	49.80	156.40	81.95	-	58.10	66.47	34.27	-	8.30
10. Udgir	16.25	5.90	-	3.95	74.75	37.76	-	47.40	130.00	64.90	-	55.30	55.25	27.14	-	7.90
District	17.83	7.16	-	3.54	82.02	45.82	-	42.48	142.64	78.76	-	49.56	60.62	32.94	-	7.08

Source: Primary Data.

above discussion, it is clear that the FPSs in the study area have fulfilled 65.17 percent of the per month requirement of wheat, 75.45 percent of the per month requirement of rice & 91.71 percent of the permonth requirement of Kerosene in the study area. However, sugar has disappeared from the list of rationed goods due to a marginal difference between its PDS retail price & open marekt price & also due to the low quality of sugar sold through FPSs. The sample families depended heavily on FPSs for meeting their kerosene requirements. This is also happened on the background of low difference between the PDS retail price and open market price of kerosene in the study area. Since there is not very much visible difference between the quality of kerosene sold through FPS and the sold in open market, the ration card holders perfer to buy kerosene from the FPS even though there is marginal difference between the PDS retail price and open market price of Kerosene. Comparing the case of sugar with Kerosene, it can be said that quality of sugar has been responsible for its disappearance from the PDS system in the study area.

The per family per month value of purchases from FPSs by the sample beneficiaries, is found to be Rs. 82.02 in case of wheat, Rs. 45.82 in case of rice and Rs. 42.48 in case of Kerosene. On the other hand, the per family permonth, Value of FPS purchases, by sample beneficiaries, at marekt prices is estmated at Rs. 142.64 for wheat, Rs. 78.76 for rice and Rs. 49.56 for kerosene, during the survey period. Hence the per family per month gain from FPS purchases is found to be Rs. 60.62 for wheat, Rs. 32.94 for rice & Rs. 7.08 for kerosene. These figures indicate that the distribution of essential commodities like rice, wheat and Kerosene through FPSs has transferred substantial gains to the sample beneficiaries of the district under study.

The taluka-wise data about gains from FPS purchases presented in table 8 reveals that the per family per months gain from wheat was the maximum in case of Shirur Anantpal (Rs. 71.40) taluka followed by Devani (Rs. 66.47) & Ausa (Rs. 66.13) talukas. It was the lowest in case of Nilanga (Rs. 54.50) taluka, followed by Ahmedpur (Rs. 58.18) & Udgir (Rs. 55.25) talukas of the district. Per family per month gains from FPS purchases of rice is found to be the maximum in case of Renapur (Rs. 34.50) taluka, followed by Devani (Rs. 34.27), Nilnaga (Rs. 34.26) & Chakur (Rs. 34.04) talukas of the district. Per family per month gain from FPS purchases of rice is found to be the lowest in case of Udgir (Rs. 27.14) & Ausa (Rs. 31.28) talukas of the district. Kerosene is distributed through FPS even though there is marginal difference between its PDS retail Price & price in the open market. The per family per month gain from PDS purchases of Kerosene accured to sample beneficiaries is estimated at Rs. 7.08. These gains from PDS purchases of Kerosene have varied in a very narrow range of Rs. 8.30 per family per month in Devani taluka & Rs. 6.70 per family per month in Ausa taluka of the district under study. From the above discussion it is clear that the sample beneficiaries of PDS have gained Rs. 60. 62 from wheat, Rs. 32.94 from rice and Rs. 7.08 from Kerosene per month during the survey period . Therefore the public distribution system is transferring considerable income to wards the weaker sections of the soceity in the study area.

REFERENCES

Bapna S.L., Role of Public Distribution System in Food Scarcity, in Agricultural Price policy for developing countries (Eds.) Mellor J.W. and Raisuddin Ahmed, 1988.

Kaur Rajbans, Agricultural Price Policy in Economic Development, Kalyani Publishers, Delhi, 1975.

Kripa Shankar, Revamped Public Distribution System-Who Benefits & How Much? Economic & Political Weekly, March 29, 1997.

Dholakia N. and Khurana R., Public Distribution System, Oxford & IBH, Delhi, 1979.

Dantawala M.L., Agricultural Policy; Price and Public Distribution System: A review, Indian Journal of Agricultural Economics, Vol. 48, No.2, 1993.

Subbarao K., Improving Nutrition in India-Policies, Programmes & their Impact, World Bank Discussion Paper 49, 1989.

Parikh K. Who Get How Much from PDS-How Effectively Does it Reach the Poor. Sarvekshana, January-March, 1994.

Dev Mahendra, Alleviating poverty-Maharashtra Employment Guarantee Scheme, Economic & Political weekly, October-14-21, 1995